Experimental Modes
of Civic Engagement
in Civic Tech

Experimental Modes
of Civic Engagement
in Civic Tech

Meeting people where they are.

LAURENELLEN MCCANN
Edited by Daniel X. O'Neil

To Sabrina, for listening, and for teaching me how to listen.

*The most common way people give up
their power is by thinking they don't have any.*
—Alice Walker

*It's a question of, how shall we live? How shall we continue the
evolution of human beings? What does it mean to be a human being at
this time on the clock of the world?*
—Grace Lee Boggs, Democracy Now!, 2011

It's people!
— Detective Thorn, Soylent Green, 1973

Experimental Modes of Civic Engagement in Civic Tech: Meeting people where they are.

by Laurenellen McCann
edited by Daniel X. O'Neil
is licensed under a Creative Commons Attribution-ShareAlike 4.0 International License.

Based on a work at http://www.smartchicagocollaborative.org/modes

Manufactured in the United States of America by the
Smart Chicago Collaborative
http://www.smartchicagocollaborative.org/
@smartchicago
c/o The Chicago Community Trust
225 North Michigan Avenue
Suite 2220
Chicago, IL 60601
(773) 960-6045

Made possible by a Knight Community Information Challenge Deep Dive grant given to The Chicago Community Trust by the John S. and James L. Knight Foundation.

Set in Scala and ScalaSans

Library of Congress Control Number: 2015953051
ISBN: 978-0-9907752-2-5

First Printing, September 2015

Contents

Preface 1

People First, Tech Second: 5 Criteria for
Community-Driven Civic Tech 3

5 Modes of Civic Engagement
in Civic Tech 8

Mode #1: Use Existing
Social Structures. 10

Mode #2: Use Existing
Tech Infrastructure 17

Mode #3: Create Two-Way
Educational Environments. 22

Mode #4: Lead From
Shared Spaces 28

Mode #5: Distribute Power 32

Convening 40

Active Listening 101 46

Real-World Civic Tech Strategies 50

Tools, Not Tech. 55

Where Does Community Organizing
End and Civic Tech Begin?. 58

Closing 64

Acknowledgements 66

Appendix: Case Study Analysis 68

Preface

Experimental Modes of Civic Engagement in Civic Tech is an investigation into what it means to build civic tech with, not for. It answers the question, "What's the difference between sentiment and action?"

The project was led by Laurenellen McCann, and it deepens her work in needs-responsive, community-driven processes for creating technology with real people and real communities for public good.

This project falls under Smart Chicago's work on the Knight Community Information Challenge grant awarded under their Engaged Communities strategy to The Chicago Community Trust "as it builds on its successful Smart Chicago Project, which is taking open government resources directly into neighborhoods through a variety of civic-minded apps."

This book is a compendium of writing by Laurenellen, originally published on the Smart Chicago blog. I'm excited about this project because it supports so many important nodes for Smart Chicago:

- Keeping the focus on people and communities rather than technology. We are leading creators of civic tech, and we publish a lot of software. It's people and impact we care about.

- Driving toward a shared language around the work. There is a lot of enthusiasm for "people" in our space right now. This project sharpens pencils and will put definition to the work.

- Highlighting the workers. Communities are doing this work and doing it right. We seek to lift them up and spread their methods.

Smart Chicago is utterly devoted to being of impact here in Chicago. As our work progresses, we see the opportunity to have influence all over. This project, rooted in The Chicago Community Trust, funded by The Trust and the Knight Foundation, executed by a leading thinker in the field, is one way we're doing just that.

— Daniel X. O'Neil, Smart Chicago Collaborative

References

"Knight Community Information Challenge." Knight Foundation. Accessed September 29, 2015. http://www.knightfoundation.org/funding-initiatives/knight-community-information-challenge.

"Communities." Knight Foundation. Accessed September 29, 2015. http://www.knightfoundation.org/what-we-fund/engaging-communities.

"Smart Chicago Collaborative." Knight Foundation. Accessed September 29, 2015. http://www.knightfoundation.org/grants/201448269.

"No More Trickle Down #CivicTech." Medium. September 30, 2014. Accessed September 29, 2015. https://medium.com/@elle_mccann/no-more-trickle-down-civictech-81341cf48a14.

People First, Tech Second: 5 Criteria for Community-Driven Civic Tech

----

What does it look like to build civic technology with, not for, the people you're trying to serve? What's the difference between "civic" sentiment and action?

That's the thrust of Experimental Modes in Civic Engagement for Civic Tech—a special initiative that I led for Smart Chicago as part of their Community Information Deep Dive. The scope of this work was guided by the "civic" in "civic technology": the real people and real communities we claim to serve when we set out to create tools for public good. Our belief was that building real civic technology, the kind that doesn't just "solve problems," but actually allows people to enhance their quality of life and (re)define their relationship with their governments, media, and each other, requires a perspective that prioritizes people above production.

Our approach had three parts: **(1) a scan of the field**, identifying practitioners of needs-responsive, community-driven tech and the basic characteristics, best practices, and models that define their work, **(2) a convening** of practitioners at The Chicago Community Trust who came together on April 4, 2015, and **(3) a book** documenting our investigation of the space as well as tactics and strategies for civic tech that refocus the work on people.

Criteria

In my initial research to inform the scan of the field, I discovered a real contrast between civic technologies that were community-driven in their approach and those that weren't.

Community-driven civic technologies are built at the speed of inclusion—the pace necessary not just to create a tool but to do so with in-depth communal input and stewardship, responding to the needs, ideas, and wants of those they're intended to benefit. In other words, they put communities in the driver's seat when it comes to identifying civic problems and crafting civic solutions. But they don't always call themselves "civic."

So to guide discovery and analysis of projects that follow this "build with, not for" approach, I developed the Criteria for People First Civic Tech, to explore the degree to which various civic tools, projects, and programs prioritize people and real world application above production.

To prioritize people and build with them is to:

1. **Start with people:** Work with the real people and real communities you are part of, represent, and/or are trying to serve

2. **Cater to context:** Leverage and operate with an informed understanding of the existing social infrastructure and sociopolitical contexts that affect your work

3. **Respond to need:** Let expressed community ideas, needs, wants, and opportunities drive problem-identification and problem-solving

4. **Build for best fit:** Develop solutions and tools that are the most useful to the community and most effectively support outcomes and meet needs

5. **Prove it:** Demonstrate and document that community needs, ideas, skills, and other contributions are substantially integrated into—and drive—the lifecycle of the project

Beyond direct application to the Experimental Modes initiative, my goal in creating these criteria was to define the leanest standard possible for translating the idea of "with" to a series of identifiable practices that can be used for further investigation, accountability, and guidance outside of this project.

Some of the principles defined above have been long mirrored in (and championed by!) the design community (and other fields), but their expression has yet to become part of the creation of mainstream civic technology today. Nor has there been much dialogue about what it means to do more than just design with a community, but to literally build and evaluate these civic tools together—to let community drive the whole process.

So, in the time we have together, that's what we'll do. Over the next few chapters, we'll walk through my research on existing practices for deep and direct community collaboration for the ideation, design, and creation of civic tools. Then, we'll take a closer look at some specific projects and hear from the people behind them about how they do what they do. At the end, we hope you'll walk away not just with new strategic insights for your work, but with some new perspective. Civic technology is a relatively young field, which means that its practitioners have a lot of opportunity to shape not just what it will become, but also who's at the table. As you read, we hope you challenge yourself to see these opportunities in the context of your practice and think critically about how you respond.

References

"Experimental Modes of Civic Engagement in Civic Tech." Smart Chicago. December 12, 2014. Accessed September 29, 2015. http://www. smartchicagocollaborative.org/work/special-initiatives/deep-dive/experimental-modes-of-civic-engagement-in-civic-tech.

"Deep Dive." Smart Chicago. January 4, 2015. Accessed September 29, 2015. http://www.smartchicagocollaborative.org/work/special-initiatives/deep-dive.

"New Project: Experimental Modes of Civic Engagement in Civic Tech." Smart Chicago. December 12, 2014. Accessed September 29, 2015. http://www.smartchicagocollaborative.org/new-project-experimental-modes-of-civic-engagement-in-civic-tech.

"BUILD WITH, NOT FOR." BUILD WITH NOT FOR. Accessed September 29, 2015. http://buildwith.org.

"Building Community And Engagement Around Data." Medium. February 23, 2015. Accessed September 29, 2015. https://medium.com/@elle_mccann/building-community-and-engagement-around-data-2fb7d72b13b4.

"A Brief History of Design Thinking: How Design Thinking Came to Be." I Think I Design. June 8, 2012. Accessed September 29, 2015. https://ithinkidesign.wordpress.com/2012/06/08/a-brief-history-of-design-thinking-how-design-thinking-came-to-be.

"Creative Process: Jeanne Marie Olson." Vimeo. Accessed September 29, 2015. http://vimeo.com/35433681.

"No More Trickle Down #CivicTech." Medium. September 30, 2014. Accessed September 29, 2015. https://medium.com/@elle_mccann/no-more-trickle-down-civictech-81341cf48a14.

"Crafting #CivicTech." Medium. November 3, 2014. Accessed September 29, 2015. https://medium.com/@elle_mccann/crafting-civictech-7f5ed-fb864bb.

Civic technology includes tools and processes we create to serve public good. It's about how we work together, which means, it's fundamentally about people. Image by Daniel X. O'Neil.

5 Modes of Civic Engagement in Civic Tech

Using the People First Criteria, I analyzed dozens of "civic technology" projects, mostly, but not exclusively within the U.S. I disregarded whether or not the projects or creators identified with "civic tech," looking instead at whether or not the "tech" in question was created to serve public good. (Our interest, after all, is to explore the "civic" in "civic tech.")

Those projects that fit the People First Criteria were diverse in terms of the technologies developed, the benefits yielded, and the communities that were (and, in some cases, still are) behind the wheel. But there are a great number of similarities, too—consistent, proven strategies and tactics that other practitioners of (and investors in) civic tech can learn from.

I classified these similarities as the "5 Modes of Civic Engagement in Civic Tech," which are listed below along with common tactics for implementation. Over the next few chapters, we'll look at each of these strategies in depth as well as case studies of some of the civic tech projects that have successfully implemented these community-driven processes for "bottom-up innovation."

5 Modes of Civic Engagement in Civic Tech

I. **Use Existing Social Structures**
    - *Pay for Organizing Capacity in Existing Community Structures*
    - *Partner With Hyperlocal Groups With Intersecting Interests*
    - *Offer Context-Sensitive Incentives for Participation*

2. **Use Existing Tech Infrastructure**
   - *Remix, Don't Reinvent*
   - *Use One Tech to Teach Another*

3. **Create Two-Way Educational Environments**
   - *Start with Digital/Media Skills Trainings*
   - *Co-Construct New Infrastructure*

4. **Lead From Shared Spaces**
   - *Leverage Existing Knowledge Bases*
   - *Leverage Common Physical Spaces*

5. **Distribute Power**
   - *Treat Volunteers as Members*
   - *Train Students to Become Teachers*
   - *White-label Your Approach*

References

"Civic Innovation Beyond Civic Technology." *New America RSS*. Accessed September 29, 2015. https://www.newamerica.org/oti/civic-innovation-beyond-civic-technology.

"No More Trickle Down #CivicTech." *Medium*. September 30, 2014. Accessed September 29, 2015. https://medium.com/@elle_mccann/no-more-trickle-down-civictech-81341cf48a14.

Mode #1: Use Existing Social Structures

Social infrastructure refers to the ecosystem of relationships and formal and informal organizations in a community. Structures can be physical (such as institutions with actual storefronts, like a day care center) or purely relational (like a parents' meet-up group), and most are organized by some element of place (neighborhood, school district, city district, city, etc.).

Although structures can be shared across communities (a day-care center can draw people from multiple neighborhoods), the particular social infrastructure of a community is always unique. One may rely heavily on the daycare center while another nearby may prefer informal babysitting co-ops or church programs. Outsiders aren't likely to spot these more informal and relational structures, making it hard to discover the structures that really matter in a given social context.

To address this knowledge gap, you need to literally meet people where they are—go to the hubs in the social network of the community you're trying to serve and work in partnership to customize an approach to technology development that best fits the community you're working with.

Here are three specific tactics, each with concrete examples of real-world use, that can help you think about using existing social infrastructure in civic tech:

TACTIC: Pay for Organizing Capacity in Existing Community Structures

Whether you're trying to catalyze new tech activity or create general opportunities for communal self-direction in tech, investing money where a community is already investing social capital is one method of working with existing social infrastructure. Investing in the capacity of organizers to expand their work and seek opportunities to leverage technology is a direct way of ensuring that tech is both situated in a communal context and won't be made an afterthought to competing priorities.

The Chicago Large Lots program has gained much attention for its tech platform LargeLots.org, which lets residents of particular neighborhoods purchase city-owned vacant lots for $1. This online platform originated not within Chicago's civic hacking community but thanks to the coordination of various neighborhood associations and community groups at the helm of the policy response to this issue. As the policy developed in coordination with the City, these groups eventually leveraged social connections to craft civic tech to help execute the policy.

These social connections existed in part due to a previous investment in organizing capacity from the federally-funded Broadband Technology Opportunities Program (BTOP). In Chicago, some BTOP money was directed toward helping local organizations hire tech organizers and digital literacy instructors to "expand digital education and training for individuals, families, and businesses."

One of those organizations was Teamwork Englewood, a community organization that would later play a role in the creation of the Large Lots program—a role that was only possible because, thanks to BTOP, Teamwork had existing paid staff whose responsibility it was to both invest in local digital skill-building and seek context-relevant opportunities to leverage those skills for neighborhood change.

Paid capacity can express itself in far more localized ways, too. For example, the student-run Hidden Valley Nature Lab, which enables teachers to modify their curricula for place-based learning using QR codes, is the product of general paid support (at both the teacher and student level) for digital educational programming within the communal social infrastructure that is New Fairfield High School, a public school in Western Connecticut.

TACTIC: Partner with Hyperlocal Groups With Intersecting Interests

Red Hook Wifi, a community-designed and stewarded wireless Internet network in Red Hook, Brooklyn, New York, is the product of a layered series of partnerships:

- A national organization, the Open Technology Institute (OTI), with expertise in community wireless networks
- A hyperlocal organization, the Red Hook Initiative (RHI), a community center devoted to social justice and restoration of local public life through youth-led approaches
- A variety of educational, residential, and local business relationships that not only utilize the network, but help expand the capacity available to keep the network alive through another RHI program, the Digital Stewards

This deep meshing of missions, skills, and structures enabled the national organization (OTI) to support hyperlocal work in a way that genuinely allowed the local organization (RHI) to not only drive, but ultimately (literally) steward the ongoing success of both the wifi network and the social infrastructure needed to keep the network relevant and present within the community.

TACTIC: Offer Context-Sensitive Incentives for Participation

Although most of the examples above focus on organizational relationships, to catalyze the participation of individuals, explore the use of specific incentives.

For example, the Civic User Testing Group (CUTGroup) is a model of user experience testing run by the Smart Chicago Collaborative that enables "regular residents" to explore and critique so-called civic apps. To date, most participants are given a $20 Visa gift card for their engagement, although Smart Chicago is exploring the use of even more contextually relevant awards, such as money for groceries for testing apps related to food access.

Sometimes getting to play a specific role in an activity can be its own currency. DiscoTechs (short for "Discovering Technology") are an open event format for teaching and sharing digital skills in a communal context and operate by utilizing social capital incentives. Although some DiscoTechs cater to specialized skills, many give neighbors, peers, and local organizations the chance to demonstrate a variety of personal technical expertise (such as photography, digital storytelling, music-making, coding, fabrication, you name it) and gain new community credibility alongside new opportunities.

Demond Drummer, former tech organizer at Teamwork Englewood, presents on LargeLots.org. Photo by Chris Whitaker.

Digital Stewards setting up the Red Hook Wifi network. Image via DigitalStewards.org.

Experimental Modes

References

No More Trickle Down #CivicTech." Medium. September 30, 2014. Accessed September 29, 2015. https://medium.com/@elle_mccann/no-more-trickle-down-civictech-81341cf48a14.

"Crafting #CivicTech." Medium. November 3, 2014. Accessed September 29, 2015. https://medium.com/@elle_mccann/crafting-civictech-7f5edfb864bb.

"Large Lots." Large Lots. Accessed September 29, 2015. http://largelots.org/about.

"BTOP." Smart Chicago. November 21, 2011. Accessed September 29, 2015. http://www.smartchicagocollaborative.org/work/broadband-technology-opportunities-program.

"Seeking Digital Literacy Instructor: Teamwork Englewood." Teamwork Englewood.

Accessed September 29, 2015. http://www.teamworkenglewood.org/news/8551.

"Hidden Valley Nature Lab." Hidden Valley Nature Lab. Accessed September 29, 2015. https://sites.google.com/site/hiddenvalleynaturelab.

"Red Hook Wifi, A Project of the Red Hook Initiative." Red Hook Wifi. Accessed September 29, 2015. http://redhookwifi.org.

"Open Technology Institute." New America Oti RSS. Accessed September 29, 2015. http://www.newamerica.org/oti.

"RHI Digital Stewards." RHI Digital Stewards. Accessed September 29, 2015. http://www.rhidigitalstewards.wordpress.com.

"CUTGroup." Smart Chicago. August 1, 2013. Accessed September 29, 2015. http://www.smartchicagocollaborative.org/work/ecosystem/civic-user-testing-group.

"The CUTGroup Book." The CUTGroup Book. Accessed September 29, 2015. http://www.cutgroupbook.org.

"Detroit Digital Justice Coalition." Detroit Digital Justice Coalition. Accessed September 29, 2015. http://detroitdjc.org/?page_id=23.

"So You Think You Want to Run a Hackathon? Think Again." Medium. June 24, 2014. Accessed September 29, 2015. https://medium.com/@elle_mccann/so-you-think-you-want-to-run-a-hackathon-think-again-f96cd7df246a.

DiscoTech stations can range from mapping and coding how-tos to digital storytelling to Scrabble, each offering unique incentives to participate for both teachers and attendees. Photo by Maureen McCann.

MODE #2: Use Existing Tech Infrastructure

"Innovations" and technologies don't have to be brand new in order to be leveraged for civic impact. Some successful tools are the product of simply using or encouraging the use of tools that communities have ready access to or already rely on in new ways. It's important to note that here, when we talk about "technical infrastructure," we're talking about both physical elements, like wireless network nodes, radio towers, and computers, and digital elements, like social media platforms, email, and blogs—the tech tools a community uses to support everyday activity and public life.

TACTIC: Remix, Don't Reinvent

Jersey Shore Hurricane News is a collaborative news "platform" built on a standard-feature Facebook Page. The site is run by Justin Auciello, who co-founded it with friends in 2011 during Hurricane Irene. Seeing the need to share hyperlocal info during the hurricane and curious about ways to use tech to spur civic engagement, as the storm took hold, Auciello decided to start a hub for emergency information that anyone could contribute to. Rather than create a separate blog or media site, Auciello went where New Jersey residents were already gathering to learn and share: Facebook.

Over time, the platform has expanded in terms of the content (it now covers real-time daily news), the role it plays in the community, and the network of volunteer contributors involved in reporting and sharing. But the tool remains the same.

This is the art of the remix: the recombination of familiar, ordinary elements to create something extraordinary.

As part of their ELECTricity project, the Center for Technology and Civic Life (CTCL) recombined familiar elements of free website templates, using Google Blogger, to help local election administrations modernize and share information. The idea of working with Google Blogger and creating a new template (rather than developing an entirely new tool) was the result of a listening tour: for nearly 7 months, the ELECTricity team discussed points of pride and pain with local election administrators around the country. After the tour, CTCL realized that, in order to support administrators' needs, the ideal online platform would need to be (almost) free and require minimal amounts of technical knowledge to get up and running while still enabling opportunities for more complex technical developments in the future. Blogger templates fit the bill.

TACTIC: Use One Tech to Teach Another

Free Geek is a non-profit organization model that was started in Portland, Oregon in 2000 and has been implemented in 12 other cities in the U.S. and Canada. Free Geek takes old computer parts and works with communities in underserved areas (particularly those with low access to digital technology) to use this e-waste to build new computers and electronics. These electronics are then made available for free—plus some community service time.

Although the Free Geek system is ultimately about making computers available to those in need, Free Geek has built out programs that leverage this tech to teach and develop additional tech skills. For example...

- Along with supplying free hardware, Free Geek supplies software and offers basic digital skills training
- Along with supplying free software, some Free Geeks also offer training in code and web design and development at a variety of skill levels

- Since community members "buy" their free computers through volunteer time, Free Geeks offer the chance to "pay" for computers by volunteering to build new electronics and trains interested community members in a variety of hardware, wiring, and refurbishing skills

By focusing on the lifecycle of technology use, Free Geek feeds into and strengthens the basic technical infrastructure of the communities it inhabits and provides platforms for partnerships and individual incentives (see Mode #1) that support public goods, like job training.

This is akin to the structure that the Red Hook Initiative (RHI) in Red Hook, Brooklyn, New York uses to manage Red Hook Wifi, a community wireless network. Red Hook Wifi is maintained through a youth program, Digital Stewards, which trains residents age 19-24 to install and maintain a wireless network that serves neighborhood homes and businesses. On top of that skill base, Digital Stewards are also taught how to do software and hardware troubleshooting as well as community organizing and public relations—skills necessary to keep the network up and running both technically and socially. Like Free Geek, the Digital Stewards program focuses on technology from an ecosystem perspective, stacking the development of new tools so that they can sustainably integrate into existing community structures.

References

"No Revolution Without Reflection." Medium. January 22, 2015. Accessed September 29, 2015. https://medium.com/@elle_mccann/ no-revolution-without-reflection-5dfa513cd43a.

"Facebook Logo." Jersey Shore Hurricane News. Accessed September 29, 2015. https://www.facebook.com/JerseyShoreHurricaneNews.

"Jersey Shore Hurricane News: Using Facebook and Crowdsourcing to Build a News Network." TechPresident. Accessed September 29, 2015. http://techpresident.com/news/25352/jersey-shore-hurricane-news-using-facebook-and-crowdsourcing-build-news-network.

"About ELECTricity." Center for Technology & Civic Life. Accessed September 29, 2015. http://www.techandciviclife.org/electricity.

"Designing A Website For Local Election Offices." ELECTricity. Accessed September 29, 2015. http://electionbuzz.tumblr.com/post/79168453775/designing-a-website-for-local-election-offices.

"Election Website Project Update." ELECTricity. Accessed September 29, 2015. http://electionbuzz.tumblr.com/post/101169668157/election-website-project-update.

Wikipedia. Accessed September 29, 2015. http://en.wikipedia.org/wiki/Free_Geek.

"We Transform Used Technology into Opportunity, Education, & Community!" Free Geek. Accessed September 29, 2015. http://www.freegeek.org/#volunteer.

"Red Hook Initiative." Accessed September 29, 2015. http://rhicenter.org.

"RHI Digital Stewards." RHI Digital Stewards. Accessed September 29, 2015. https://rhidigitalstewards.wordpress.com.

Portland volunteers at work. Image by Free Geek.

MODE #3: Create Two-Way Educational Environments

The first two modes encompass strategies and tactics for starting civic technology projects within existing community contexts, both in terms of social infrastructure and technical infrastructure. The next three modes will address ways to affect these structures.

Adding new technology into the infrastructure of a community is more complicated than simply teaching community members how to use the new tech. For the skills and tech-use to stick, communities must have the opportunity to integrate the new tools and new skills into their lives, on their own terms. In an educational setting, this translates to allowing community members to tinker—to play, feel ownership, and figure out how they relate to the tech (or don't).

It also means creating environments where the teacher is actively listening and responding to the ideas and sticking-points offered by participants. Rather than pushing on the development of a single skill, a teacher in a two-way educational environment treats every training as an opportunity to listen as well as be heard.

As people learn, they tend to express wants and needs that are particular to the tool they're using as well as how that tool could relate to their lives. Two-way teachers keep their ears perked for both, and seize opportunities where issue overlap allows for skills development to translate into community-driven tech development.

TACTIC: Start With Digital/Media Skills Training
Many community-driven civic technologies are the product of training in foundational media and digital skills that open up immediate and long-term opportunities for co-development.

Hidden Valley Nature Lab, a student-run experiment in place-based learning using QR codes, came from digital skills training at a public high school. Teachers gave students the chance to develop an idea inspired, but not directly taught, during class and the lab and associated work developed as a direct result.

The impact of digital training is not always so immediate, however. LargeLots.org, the web platform for purchasing city-owned vacant lots discussed briefly in the last chapter, was made possible through paid support for a digital literacy instructor ("tech organizer") at a local community organization. This trainer's job was explicitly to teach, listen, and find opportunities to connect the needs of community groups with appropriate technological solutions—something residents were able to capitalize on during the development of the Large Lots Program.

Similarly, The NannyVan's Domestic Worker Alliance App was the product of a long-tail two-way educational initiative. The "app" is actually just a phone hotline structured around a fictional, educational show. It was developed in coordination with a hyperlocal partner and the domestic worker community in New York. The NannyVan developed a relationship with this community through a media production training. Later, when an advocacy opportunity arose, the local partner turned to The NannyVan team to co-develop a tool that would best fit their social and technical needs, trusting The NannyVan's approach based on their previous experience.

This tactic is probably best seen, however, in the creation of Detroit Future Media, an intensive digital literacy program crafted to support Detroit's revitalization, created by the Detroit Digital Justice Coalition. In 2009, fueled by a grant from the Broadband Technology Opportunities Program (BTOP), the Allied Media Projects (AMP) had an opportunity to expand broadband Internet adoption in Detroit's underserved communities—communities that were already reaching out to AMP looking for digital and media skills trainings.

As AMP notes in a later report, when they approached the idea of expanding not just how the Internet could be physically accessed, but how digital technologies could be sustainably leveraged by communities for their own needs, they encountered an unavoidable capacity gap.

"...there were few people in [Detroit] who had the special combinations of technical skill, teaching experience in non-academic settings, community connectedness and desire to use media for community revitalization."

So, AMP had an idea: what if the BTOP grant could be used to train trainers —folks who were already acting as teachers, connectors, and leaders in the context of Detroit's many communities? To pull this off, AMP joined with 12 other community organizations to create the Detroit Digital Justice Coalition, which applied for BTOP funds to create the Detroit Future Media trainings along with a few other programs.

Approaching technology training from this relational perspective allowed the impact of teaching one individual to be immediately amplified and interconnected through social infrastructure—and created new structures that support continued development at both hyperlocal and city-wide scales. One outcome was the creation of a Digital Stewards program to create and maintain community wireless networks across Detroit.

TACTIC: Co-Construct New Infrastructure

Methods of teaching while listening are not only effective for sharing skills, but also building tools. Between 2002 and 2010, the Prometheus Radio Project worked with over 12 communities around the world on barnraisings—a method of rapid construction for community radio stations, with an explicit nod to the Amish tradition. Each radio barnraising brings together residents through the stewardship and organizing work of a local community group (see Mode #1) and radio experts and advocates from around the region to get a station from zero to live on the air over the course of three days. In addition to literally co-creating new technical infrastructure, volunteer facilitators lead workshops throughout the barnraising to get community members up to speed on federal regulation, radio engineering, programming and the lobbying and advocacy needed to keep their stations on the air.

Although Prometheus aided in the format of the event and the literal construction, in every instance, the education and development that occurred over the course of the barnraising was shaped by the input of the convening community group and all the participants in the event.

Installing new technical infrastructure through collaborative educational processes that instill community ownership is also readily present in the work of:

- Red Hook Wifi: a community wireless network in Brooklyn that is maintained by the Digital Stewards, an educational program for young adults

- Free Geek: which provides access to free computers built by a community for a community

- Public Lab: an international community of citizen scientists who develop and share tools and techniques to aid in each other's distributed research

References

"Hidden Valley Nature Lab." Hidden Valley Nature Lab. Accessed September 29, 2015. https://sites.google.com/site/hiddenvalleynaturelab.

"Large Lots." Large Lots. Accessed September 29, 2015. http://largelots.org/about.

"NannyVan." NannyVan. Accessed September 29, 2015. http://www.nannyvan.org.

"The Domestic Worker App." Accessed September 29, 2015. https://vimeo.com/84907598.

"Detroit Digital Justice Coalition." Detroit Digital Justice Coalition. Accessed September 29, 2015. http://detroitdjc.org.

"Allied Media Projects." Allied Media Projects | Media Strategies for a More Just and Creative World. Accessed September 29, 2015. http://alliedmedia.org.

"Detroit Future Media Guide to Digital Literacy." Accessed September 29, 2015. https://www.alliedmedia.org/files/dfm_final_web.pdf.

"Detroit Digital Stewards." Detroit Digital Stewards. Accessed September 29, 2015. http://detroitdigitalstewards.tumblr.com.

"Prometheus Radio Project." Prometheus Radio Project. Accessed September 29, 2015. http://www.prometheusradio.org.

"Barnraisings." Barnraisings. Accessed September 29, 2015. http://www.prometheusradio.org/barnraisings.

"We Transform Used Technology into Opportunity, Education, & Community!" Free Geek – Computer/Technology Reuse, Education, Recycling & Sales. Accessed September 29, 2015. http://www.freegeek. org/#volunteer.

"Public Lab Is a Community Where You Can Learn How to Investigate Environmental Concerns." Public Lab: A DIY Environmental Science Community. Accessed September 29, 2015. http://publiclab.org.

A crowd waits for a world premiere broadcast after the Pineros y Campesinos Unidos del Noroeste Barnraising in Woodburn, Oregon. Image by The Prometheus Radio Project.

MODE #4: Lead from Shared Spaces

Communities are built around commons—collaboratively owned and maintained spaces that people use for sharing, learning, and hanging out. Commons are the foundation upon which all community infrastructure (social, technical, etc.) is built and are often leveraged by multiple overlapping and independent communities.

Although often thought of as permanent physical spaces, like parks or town centers, commons can also be digital (i.e. online forums, email lists, and wikis), temporary (like pop-ups or weekend flea markets), or a variety of other set-ups beyond and in-between.

A commons is a resource, offering tools, news, and know-how that both community insiders and outsiders can wield. Tapping into a commons not only helps identify social and technical infrastructure, it provides a key opportunity to listen and learn about what matters most to a community. The following two tactics look at how civic tech practitioners can not only use commons for collaborative work, but can also contribute to their stewardship.

TACTIC: Leverage Existing Knowledge Bases

Knowledge commons are spaces where people collect and access information, be it archival info (like one would get from a library) or news (like one gets from a neighborhood listserv).

Depending on the circumstances and the folks behind the wheel, the creation of a knowledge commons can itself be a form of civic technology—a tool for a community to use for its own benefit.

DavisWiki is a hub for both hyperlocal history and current events in Davis, California. Launched in 2004, DavisWiki started as an experiment in collaboratively surfacing and capturing unique local knowledge that was otherwise locked in the heads of neighbors or lost in search engines. The site gained popularity by coordinating with existing social infrastructure, such as the university system in Davis and the local business community, and within a few years residents had contributed over 17,000 pages.

As more residents use DavisWiki, the platform's role has changed. In addition to being a popular catalog, knowing that DavisWiki is available as a knowledge commons has enabled residents to leverage the platform for additional civic ends over time. For example, the wiki was part of the coordinated public response to a police officer pepper-spraying a student on the UC Davis campus in 2011 and has been used to explore, discuss, and collaborate with government on a number of local planning initiatives before and since.

Public Lab is a knowledge-based community focused on environmental science. Although many of the folks who participate in Public Lab (via their wiki, email listservs, in-person meetings, and other forums) do hyperlocal work, the community associated with Public Lab is international in scope, bringing together citizen scientists from around the globe who are researching and crafting inexpensive DIY tools for environmental monitoring.

Although there are plenty of other issue-focused wikis, professional networks, and meet-up groups out there, Public Lab remains distinct because of its emphasis on connection. First, it focuses on the connection of disparate pieces of local knowledge, allowing community members to integrate their learnings into a public-facing resource that others can learn from and add to. Second, it focuses on the real human connections of its members, devoting paid organizing capacity (see Mode #1) and volunteer resources (see Mode #5) to inviting and integrating relationship-building into all parts of its work. The result is a complex, but collaborative system of civic tool creation and information stewardship that focuses on (and always returns to) people.

TACTIC: Leverage Common Physical Spaces

Although the network model of Public Lab enables a high degree of exposure for local work, nothing says "free PR" quite like door-knocking or standing on your neighbor's roof to install an Internet router. Both the Detroit and Red Hook Digital Stewards programs lead with this idea, leveraging common spaces in their communities (neighborhoods and city districts) to plug into existing social infrastructure and invite community members to learn more and get involved.

Approaching technology development with an eye toward the physical world also enables an additional dimension of sustainability. While both Digital Stewards programs are ultimately about developing digital commons, by tapping into physical resources and the social structures that maintain them, the Stewards extend communal care and maintenance to include the new technology over time.

References

"Davis LocalWiki." Davis. Accessed September 29, 2015. https://localwiki.org/davis.

"Wiki History." Davis. Accessed September 29, 2015. https://localwiki. org/davis/Wiki_History.

"LocalWiki for Civic Engagement." LocalWiki for Civic Engagement. Accessed September 29, 2015.

https://localwiki.org/main/LocalWiki_for_Civic_Engagement.

"Public Lab Is a Community Where You Can Learn How to Investigate Environmental Concerns." Public Lab: A DIY Environmental Science Community. Accessed September 29, 2015. http://publiclab.org.

Community meeting. Image via Public Lab.

MODE #5: Distribute Power

The art of leading a collaborative process is the art of getting out of the way. You can follow best practices—building your work through public commons, rooting your projects in the existing social and technical practices of a community, and teaching new technical skills while listening—but if you can't get out of the way, you can't run a community-driven development process.

Getting out of the way means sharing project control with the group. Below are four essential tactics for sharing and releasing power that have been applied in the creation of civic technology.

TACTIC: Treat Volunteers as Members

Another title for this tactic could be: value your participants equally. Put everyone on the same level. No matter their status—whether a person contributes once to a project or 50 times, whether they lead a process or follow, whether they're paid staff or high school students—treat the folks who participate in your project as equals. For example:

- Jersey Shore Hurricane News, a Facebook-based journalism outlet, considers anyone who submits a photo, an event, a story, or a tip to be a "contributor."

- Although technically a non-profit, Public Lab, the DIY citizen science group, identifies all participants (folks who contribute to the listservs, wiki, in-person events around the world, etc.) to be part of Public Lab itself and has created structures (like working groups) for community input on decision-making.

- Free Geek is a non-profit that works with communities to transform old technology into new electronics made available to those in need. Free Geek runs on human power and uses community service as a currency. However, it makes little distinction between roles as all are valuable. So, whether you're learning how to refurbish technology, building computers in the shop, teaching a class, sorting donations, or helping to keep the facility clean, you're a "volunteer," and you are essential.

This tactic is a foot-in-the-door technique to build trust, and it's and a way of demonstrating that any contributions a person offers as part of the co-development process will be valued—an important message to send if you want actually want a diverse group of community members to feel invested and free to drive a project.

TACTIC: Teach Students to Become Teachers

Handing off control and treating people as equals doesn't mean removing structure or leadership. Projects that sustain community development are those that enable participants to expand their skills and responsibilities as they're interested in doing so, with specific tracks for leadership that are accessible to everyone from the onset. For example:

- As noted above, by default, all participants in Public Lab are identified as part of Public Lab. But for those participants who are interested or active in coordinating projects or contributing to the Lab at a different scale (from helping with communications to moderating community discussion lists), Public Lab has an open call for community leaders, called "organizers," which anyone can join.

- Digital Stewards programs often enable graduates to mentor, if not fully teach, the next group of stewards, further developing the technical skills individuals pick up from the program and deepening the communal history and relationship with the wireless networks the stewards oversee.

- Mobile Voices (or VozMob) is a content and technology-creation platform built for, by, and with, immigrant and low-wage workers in Los Angeles. VozMob has a few mechanisms for individual and group-level leadership, including a tier of "Affiliates," peer organizations and groups who are active in sharing stories through the platform who take part in decision-making.

TACTIC: White-label Your Approach

"White-labeling" means putting a product, service, or program model out into the world in such a way that anyone can rebrand it as though they made it.

For example, several times throughout this series we've looked at a program called "Digital Stewards" in both Detroit and Red Hook, Brooklyn. Although these two programs use the same language ("Digital Stewards") in reference to a training program to help design, build, and maintain community wireless networks, the programs are not one and the same, nor are they "chapters" or the expression of a single brand.

Rather, each Digital Stewards program is an imprint of a white-labeled training course on digital stewardship developed by the Open Technology Institute (OTI) at New America and the Allied Media Projects (AMP) that is available for anyone to adopt and use. OTI identifies its role in the creation of Digital Stewards as a "resource center," adding to the Digital Stewardship materials over time and responding to requests from communities (like Red Hook, Brooklyn) for support in training. Neither OTI nor AMP exerts copyright or brand control over how the program exists in the world and neither identifies as "owning" the program.

Removing "ownership" is a direct expression of the open ethos that drives civic tech and is a way of ensuring that communities have the opportunity to exert genuine ownership over a technology (or other civic project), even if the development of this project is guided by an external organization.

Other examples:

- DiscoTechs (short for "Discovering Technology") are a model of collaborative events for creating and exploring community technologies. The DiscoTech model was developed by the Detroit Digital Justice Coalition (DDJC), but DiscoTechs maintain no branding affiliation or ties to DDJC, allowing communities all over the country and the world to customize, remix, and implement as they see fit.

- The CUTGroup ("CUT" is short for "Civic User Testing") was originally designed by Smart Chicago Collaborative as a way for residents in Chicago to use civic apps and give feedback to developers. In 2014, Smart Chicago released the nuts and bolts of the program as an online guide so that others could use and riff off the model. Note that the "CUT" branding is not specific to Chicago or Smart Chicago and that the model is available for use without explicit consent from or identification with Smart Chicago.

TACTIC: Be a Participant

Ultimately, the best way to build civic tools with, not for a community is to be part of that community, sharing experiences as a member and engaging in the wants, needs, and interests of not abstract "people," but of your friends, neighbors, and colleagues.

This kind of participation is a mindset shift. Instead of leading, you are listening. (Much like the two-way teaching style discussed in Mode #3.) Whether you're an individual doing work close to home or an organization supporting distant activity, to be a participant is to allot time and space to others, and to show up for them before asking them to show up for you.

- Laura Amico, a crime reporter and co-creator of Homicide-Watch, a platform for following murder cases in Washington, D.C., struggled to find information about the murder cases she cared about. After watching her neighbors and observing how victims' and suspects' families were haphazardly monitoring information on an individual level, Laura and her husband, Chris, began to design a platform that would allow for collaborative coverage, with more data sources and opportunities for communication. HomicideWatch is the product of innovation, yes, but also shared grief and shared struggle.

- Public Lab's paid staff directly coordinates with and wields a number of communications platforms to listen to its extended community and brings together its network in an annual meeting (called a "barnraising") to build relationships, tinker with tech, make big decisions, and break bread.

- EPANow is an ongoing experiment in youth-driven hyper-local news co-founded by Stanford University Knight Journalism Fellow Jeremy Hay with residents of East Palo Alto, California. Jeremy is not an East Palo Alto (EPA) native, but is helping to steward the project after local community activists asked for his help. Hay started with a defined vision of what the news platform would be, but has since slowed his approach, both directly in response to community challenge and in response to his own revelations and experiences working with (and, increasingly as part of) the EPA community."

While I am not superfluous to the process, and what I bring to it is important, I am of necessity secondary." –Jeremy Hay

(More about Jeremy's journey participating in EPA as an outsider is documented here: http://knight.stanford.edu/life-fellow/2015/fellow-stepped-up-to-help-local-news-site-and-learned-to-step-aside)

References

"Jersey Shore Hurricane News." *Jersey Shore Hurricane News*. Accessed September 29, 2015. https://www.facebook.com/JerseyShoreHurricane-News.

"Jersey Shore Hurricane News: Using Facebook and Crowdsourcing to Build a News Network." *TechPresident*. Accessed September 29, 2015. http://techpresident.com/news/25352/jersey-shore-hurricane-news-using-facebook-and-crowdsourcing-build-news-network.

"Public Lab." *Public Lab: A DIY Environmental Science Community*. Accessed September 29, 2015. http://publiclab.org.

"Public Lab Wiki Documentation." *Public Lab*. Accessed September 29, 2015. http://publiclab.org/about.

"We Transform Used Technology into Opportunity, Education, & Community!" *Free Geek*. Accessed September 29, 2015. http://www.freegeek.org.

"We Transform Used Technology into Opportunity, Education, & Community!" *Free Geek*. Accessed September 29, 2015. http://www.freegeek.org/#volunteer.

Wikipedia. Accessed September 29, 2015. http://en.wikipedia.org/wiki/Foot-in-the-door_technique.

"Public Lab Wiki Documentation." *Public Lab*. Accessed September 29, 2015. http://publiclab.org/wiki/organizers.

"Digital Stewards." *Digital Stewards*. Accessed September 29, 2015. http://digitalstewards.org.

"VozMob." *VozMob*. Accessed September 29, 2015. http://vozmob.net.

"Participation." *VozMob*. Accessed September 29, 2015. http://vozmob.net/en/participation.

"About Digital Stewardship." Digital Stewards. Accessed September 29, 2015. http://digitalstewards.org/about.

"Open Technology Institute." New America. Accessed September 29, 2015. http://www.newamerica.org/oti.

"New America." New America. Accessed September 29, 2015. http://newamerica.org.

"Allied Media Projects." Allied Media Projects. Accessed September 29, 2015. http://alliedmedia.org.

"Detroit Digital Justice Coalition." Detroit Digital Justice Coalition. Accessed September 29, 2015. http://detroitdjc.org/?page_id=23.

"Detroit Digital Justice Coalition." Detroit Digital Justice Coalition. Accessed September 29, 2015. http://detroitdjc.org.

"Codesign." Codesign. Accessed September 29, 2015. http://codesign.mit.edu/discotechs.

"So You Think You Want to Run a Hackathon? Think Again." Medium. June 24, 2014. Accessed September 29, 2015. https://medium.com/@elle_mccann/so-you-think-you-want-to-run-a-hackathon-think-again-f96cd7df246a.

"Civic User Testing Group." Smart Chicago Collaborative. Accessed September 29, 2015. http://www.cutgroup.org.

"The CUTGroup Book." The CUTGroup Book. Accessed September 29, 2015.

"Homicide Watch D.C." Homicide Watch D.C. Accessed September 29, 2015. http://homicidewatch.org.

"What running Homicide Watch has taught me about crime in America." Accessed September 29, 2015. http://www.theguardian.com/cities/2014/jun/26/homicide-watch-crime-america-victims-suspects-justice.

"The Memory Keeper: Homicide Watch DC." Washingtonian. February 10, 2012. Accessed September 29, 2015. http://www.washingtonian.com/articles/people/the-memory-keeper-homicide-watch-dc.

"Public Lab Wiki Documentation." Public Lab: Barnraising. Accessed September 29, 2015. http://publiclab.org/wiki/barnraising.

"EPANow." EPANow. Accessed September 29, 2015. http://epanow.us.

"Fellow Helping to Start Local News Site Learns to Step aside." John S Knight Journalism Fellowships at Stanford. Accessed September 29, 2015. http://knight.stanford.edu/life-fellow/2015/fellow-stepped-up-to-help-local-news-site-and-learned-to-step-aside.

"Building Journalism With Community Starts With Building Trust." The Local News Lab. March 11, 2015. Accessed September 29, 2015. http://localnewslab.org/2015/03/11/building-journalism-with-community-starts-with-building-trust.

Convening

On April 4, 2015, as part of the Experimental Modes project, we
gathered 30 technology practitioners in a one-day convening at The
Chicago Community Trust to discuss the strategies they use to make
civic tech—though very few attendees would call their work such.
Artists, journalists, developers, moms, community organizers, students, entrepreneurs (and often, some combination of the
above), the practitioners in the room represented diverse parts of
the civic ecosystem and the words we each used to talk about the
work that we do reflected that.

Below, we've rounded up thoughts from each participant in answer to the question:

**Before you came into this room, did you think of your work as "civic
tech"? If you didn't, how would you describe your work?**
The answers provide an important window into the limits and
potentials of "civic technology": who feels invited into this latest
iteration of the "tech for good" space and who doesn't (or who
rejects it) and why. What follows is a slightly edited transcript of the
conversation that evolved in response to the prompt. You can find
the full, unedited notes at http://bit.ly/modesconvening.

Marisa Jahn (The NannyVan App): At first we called our work public
art, but then we identified as civic tech because the White House
called us.

Maegan Ortiz (Mobile Voices): I identified the work as civic tech because I was told that what I do is civic tech, though with the populations I work with, civic engagement has a particular meaning.

Geoff Hing (Chicago Tribune): If you owned the language, what language would you use to describe your work?

Maegan Ortiz: Great question—for me, we have meetings and make media. We're putting ourselves out there in different ways.

Marisa Jahn: We code switch a lot. Communications, civic media.

Asiaha Butler (Large Lots Program): We're open to being as "Googleicious" as possible. What we do is community.

Geoff Hing: I call my work journalism/journalistic.

Greta Byrum (Open Technology Institute): "Training."

Stefanie Milovic (Hidden Valley Nature Lab): I'd call it "civic tech". The only people who get involved are people who are looking to learn.

Jeremy Hay (EPANow): I'd call it civic tech depending on the grant. Otherwise, "community journalism."

Tiana Epps-Johnson (Center for Technology and Civic Life): Skills training and civic tech.

Naheem Morris (Red Hook Digital Stewards): Training.

Laura Walker McDonald (SIMLab): For FrontlineSMS, I'd say m-gov, m-health, etc. Digital Diplomacy. Civic tech. But the term I like the most is "inclusive technology," which baffles people because we made it up.

Robert Smith (Red Hook Digital Stewards): Training, skill building. Not tied into government, so "civic" may not apply. Community building. "Independent." Tied in to building the Red Hook community.

Jennifer Brandel (Curious Nation): Well, now I'm going to start using "civic tech" for grants. Usually, though, we call our work "public-powered journalism." Sometimes I think about our work in terms of psychogeography: "a whole toy box full of playful, inventive strategies for exploring cities ... just about anything that takes pedestrians off their predictable paths and jolts them into a new awareness of the urban landscape." ("A New Way of Walking")

Demond Drummer (Large Lots): I only started using "civic tech" about 6 months ago. Usually I refer to the work of tech organizers as "digital literacy" and "digital leadership," in the mode of the literacy trainings from the Mississippi Freedom Movement. Now I think of what I do as the "full stack of civic tech."

Josh Kalov (Smart Chicago Collaborative): Open data and website stuff. Everything I do is civic tech though I hate the term.

Anca Matioc (AbreLatAm): I work with a foundation in Chile, similar to Sunlight Foundation. Building platforms to inform people about voting, political issues. I hate the term "civic tech." It's missing a lot of what you guys in the room have, missing the communities part, the engaging grassroots part. People from civic tech need more of that. I'm impressed with R.A.G.E. (Asiaha's organization), their structure and constituent funding (and therefore their constituent accountability). Maybe that's why organizations like R.A.G.E. don't immediately identify as civic tech, because they don't have to adopt language for funders.

Allan Gomez (The Prometheus Radio Project): I don't use the term civic tech, but our work does fall under it. I'd call it "participatory democracy." Having a voice (through radio) is a civic ambition. Electoral politics is not the full range of civic participation. What about non-citizens? People who don't vote can be politically engaged in a really deep way, more so than people who only vote and that's it.

Sanjay Jolly (The Prometheus Radio Project): Our work falls into civic technology frames—and that can be important, useful. For a long time Prometheus was a "media justice organization" (to tell funders "what we are"). Now nobody call themselves media justice anymore. What makes sense to people is to say that we're building a radio station so people can have a voice in their community.

Whitney May (Center for Technology and Civic Life): Our work fits pretty squarely with civic tech language because we're building tools for government. But it's also skills training, so I'd also call it "technically civic."

Sabrina Raaf (School of Art and Design at the University of Illinois at Chicago): I'd call it open source culture. Documenting new tech. Teaching new tech. Bridging between academia and maker culture (two cultures that are biased against each other). "Sharing knowledge," documenting knowledge, workshopping knowledge.

Daniel X. O'Neil (Smart Chicago Collaborative): I work in civic tech, and I find the people in civic tech deeply boring.

Sonja Marziano (Civic User Testing Group, Smart Chicago Collaborative): "Civic" is a really important word to what I do every day.

Maritza Bandera (On The Table, The Chicago Community Trust): I never thought of what I did as "civic tech" before. Conversation. Community-building. Organizing.

Adam Horowitz (U.S. Department of Arts and Culture): Social imagination, cultural organizing, building connective tissue in social fabric.

Danielle Coates-Connor (GoBoston2030): Something I haven't seen in the civic tech space is about the interior condition of leaders...the visionary elements.

Diana Nucera (Allied Media Projects): I think of civic tech more as product than process. It's hard to hear people wanting to take the term and use it because it takes several processes to create a product that can scale to the size of civic tech—beyond a neighborhood, something that can cover a whole area. Taking over the term civic tech de-legitimizes the history of social organizing. When we use blanket terms we have to start from scratch. What I do is "media-based organizing." The work is heavy in process, not products. The products are civic tech. So, I discourage people from using the words civic technology to get grants, and so on. We actually need more diversity in processes—that's what can make civic tech valuable.

Laurenellen McCann (Smart Chicago Collaborative): This is something I've been struggling with as I've been doing my research— it's a study of processes people use to create civic tech ... but I've been wrestling with whether and how things that self-identify as "civic tech" count.

Diana Nucera: What you've shown us is that community organizing, media making, public art, all have a place within civic tech. And what I find helpful is to understand how people are approaching it: "civic tech" or "community tech."

References

"Experimental Modes Convening -- Meeting Notes." Google Docs. Accessed September 29, 2015. http://bit.ly/modesconvening.

Wikipedia. Accessed September 29, 2015. http://en.wikipedia.org/wiki/ Psychogeography.

Wikipedia. Accessed September 29, 2015. http://en.wikipedia.org/wiki/ Pedestrians.

Wikipedia. Accessed September 29, 2015. http://en.wikipedia.org/wiki/ Awareness.

"A New Way of Walking." Utne. Accessed September 29, 2015. http:// www.utne.com/community/a-new-way-of-walking.aspx.

Wikipedia. Accessed September 29, 2015. http://en.wikipedia.org/wiki/ Freedom_Summer.

"R.A.G.E." RAGE. Accessed September 29, 2015. http://ragenglewood. org.

"@Mrs_Englwood." Accessed September 29, 2015. https://twitter.com/ mrs_englewood.

Attendees of the Experimental Modes Convening. April 4, 2015.
Photo by Daniel X. O'Neil.

Active Listening 101

"We begin by listening."

This is one of the foundational principles guiding the work and communal activity of the Allied Media Projects.

Listening, active listening, is the art of focusing: lending your full attention to what a person or a bunch of people have to say and how they say it before responding. Both a skill and a practice, listening is the key to collaboration. How can you act with another person, in good faith, on equal terms, if you never take the time to focus on them, hear them, and reflect together?

When we take action and make decisions from a place of mutual understanding, a place we attain by listening and hearing, we can do more than just respond to issues. We can strengthen the essential social fabric that makes a community a community, capable not just of "surviving problems" but taking care of itself and thriving.

Listening is an essential civic skill.

At the Experimental Modes convening in April, we flexed our listening muscles as a way to learn more about each other's work, translate the jargon of our various professional backgrounds and personal experiences, and explore how we tell stories about the work that we do. This exercise was also used as preparation for something we called the case study sprint—a collaborative documentation project inspired by BookSprints. (You can learn more about this project and its results in the Appendix.)

Below, I've outlined the exercise we used. Try it out with a partner and try discussing your own work. What do you learn? What sticks with people when you tell your story that you didn't expect? What do you spend time explaining? What resonates and what doesn't?

Sample Active Listening Exercise
3–5 minutes: Define the goal and the rules
The goal of this exercise is to learn how to understand your civic project through the eyes of someone less intimately acquainted with what it is and why you do it. You'll work in pairs, taking turns playing the role of Speaker and Listener. When it's your turn to speak, you speak. When it's your turn to listen, you goal is to be as quiet as possible, focusing on what the other person says. You can take notes or doodle or whatever helps you listen, but your role is to listen. That means, no questions, no clarifications, no corrections. Just focus.

3 minutes: Break into pairs (or grab a buddy)
Pick an "A" and a "B". Person A will be the first Speaker. Person B will start as the Listener.

Round 1—Discuss WHAT You Do
2 minutes: Person A tells their story
For the next 2 minutes, Person A will be the Speaker, explaining who what their project is, who it's for/who is involved, and, briefly, how it got started. Person B is the Listener. (Again, offering no feedback or solicitation.)

3 minutes: Person B tells Person A's story.
For the next 3 minutes, Person B will reflect back what they heard from Person A. Person B speaks. Person A is only allowed to listen (and take notes)—no corrections.

5 minutes: Group reflection.
(Scaled to as many participants are part of the exercise.) What's hard about this exercise? What did Person B hear that you (Person A) didn't expect? What did they miss? What words did they use that you don't normally use to describe your work?

Round 2—Switch!
2 minutes: Person B tells their story
3 minutes: Person A tells Person B's story
2 minutes: Reflection

Round 3—Discuss HOW You Do What You Do
2 minutes: Person A shares how
Person A is the Speaker. Person B is the Listener. Talk about how the project works, how you go about implementing it and connectiong with the people you're working for, etc.—the strategic and tactical bits you'd share if you were making a recipe of your work. While you (Person A) talk, Person B will listen, taking notes on the key strategies and tactics they hear.

2 minutes: Person B shares how
Switch! Person B is the Speaker. Person A is the Listener. While Person B shares their recipe, Person A takes notes.

5 – 7 minutes: Reflection and review
Together, look through the strategies and tactics identified. Think about and discuss the language your partner used to capture your description of how you do your work. What stands out? Take another few minutes and review what's similar and different about the way you approach your work. If you're doing this exercise in a group setting, after some internal conversation time, open this topic up to the group.

What do you see about your work that you didn't see before? Which parts resonated—or didn't? How will that affect your storytelling going forward?

References

"Network Principles." AMP Wireframe. Accessed September 29, 2015. *https://www.alliedmedia.org/about/network-principles.*

"Show Your Work: Submit a Civic Tech Case Study." Smart Chicago. April 10, 2015. *Accessed September 29, 2015. http://www.smartchicago-collaborative.org/show-your-work-submit-a-civic-tech-case-study.*

Active listening: the art of civic reflection. Photo by Laurenellen McCann.

Real-World Civic Tech Strategies

At the convening, we spent some time discussing and reflecting on the 5 Modes of Civic Engagement in Civic Tech and identifying points of resonance and dissonance in our respective projects as a group. Then we split into pairs for more focused exchange. At the end, each pair was asked to reflect back to the group what strategies and tactics they found present in both their projects or what made finding commonalities difficult.

What follows are the results of our share-out. The comments have been slightly edited for formatting and clarity and annotated, when appropriate, with corresponding Modes of Civic Engagement in Civic Tech. You can read the raw meeting notes of the share-out at http://bit.ly/modesconvening.

Taken together, this forms a picture of the lack of one-size-fits-all in civic tech.

Marisa Jahn (of the The NannyVan App) and Tiana Epps-Johnson (of ELECTricity):

- Work in partnership with local organizations & working with local elections
  - *See Mode #1: Use Existing Social Structures*
- Offer specific and discrete tools and skills to groups that they need first to build trust; then moving the dialogue towards exploring shared space of collaboration
  - *See Mode #4: Lead From Shared Spaces*
- Get on the road (literally meet people where they are) to listen to your community's stories

Anca Matioc (of AbreLatAm) and Josh Kalov (of Smart Chicago Collaborative):

- Hard to draw commonalities between our work
- One overlap: both work on evolving platforms that enable multiple communities (in Latin America and Cook County, respectively) to communicate with each other
  - *See Mode #4: Lead From Shared Spaces*

Laura Walker McDonald (of SIMLab) and Geoff Hing (of Chicago Tribune):

- Didn't have commonalities. Work is done at very different orientations. Geoff's work as a code writer versus Laura's work coordinating stakeholders around technology
- Geoff drew images to show the tangle of networks each works in and they both found the people left out of that tangle tend to be the community (the people you're serving): they are not necessarily the people who are raising the funds and having to produce "outcomes" or the bottom line

Robert Smith (of Red Hook Digital Stewards) and Sanjay Jolly (of The Prometheus Radio Project):

- Similar approaches: peer-to-peer communication (reaching out to local, trusted leaders in a community that knew others)
  - *See Mode #1: Use Existing Social Structures*
- Shared values and vision between our work: the importance of local control and knowledge of technology infrastructure
  - *See Mode #2: Use Existing Tech Infrastructure*

Jennifer Brandel (of Curious Nation) and Danielle Coates-Connor (of GoBoston2030):

- The projects we're both doing meaningfully involve reaching out to community stakeholders
  - *See Mode #1: Use Existing Social Structures*
- Both collaborate with highly bureaucratic organizations with implicit norms and they (as practitioners) are trying to influence these norms and shift the dominant narrative
- Both use the frame of curiosity and questions as part of our approach to engagement
  - *See Mode #4: Lead From Shared Spaces*

Demond Drummer (of Large Lots) and Maegan Ortiz (of Mobile Voices):

- Both projects are "narrative-aware"—sensitive to what it means to work with communities in the creation of tech versus working on their behalf
  - *See Mode #5: Distribute Power*
- Focus on not creating new work but leveraging and appropriating existing technology to support existing leaders/work
  - *See Mode #2: Use Existing Tech Infrastructure*

Asiaha Butler (of Large Lots) and Allan Gomez (of The Prometheus Radio Project):

- A lot of parallels including a focus on using policy change as part of organizing but not the end-goal
- The goal is ultimately the process of media empowerment and civic engagement, with indigenous people from a shared space pushing and creating and shaping policy
  - *See Mode #3: Create Two-Way Educational Environments*
  - *See Mode #4: Lead From Shared Spaces*
- Another commonality: we were both fighting a "Goliath"

Sabrina Raaf (of University of Illinois at Chicago) and Sonja Marziano (of CUTGroup/the Smart Chicago Collaborative):

- Both bridge gaps between communities by listening to the public and having community constituencies talk to developers directly to inform how applications and tech are shaped
- Act as a conduit
- *See Mode #3: Create Two-Way Educational Environments*

Maritza Bandera (of On the Table) and Whitney May (of ELECTricity):

- Both scale work by customizing projects (via toolkits and simple templates) specific to a group while working within organizational guidelines
- *See Mode #5: Distribute Power*

Adam Horowitz (of U.S. Department of Arts and Culture) and Diana Nucera (of Allied Media Projects):

- Work with people of multiple generations
- *See Mode #1: Use Existing Social Structures*
- Use iterative processes of listening
- *See Mode #3: Create Two-Way Educational Environments*
- Use play as a vehicle for building critical connections and as an invitation to get people to the table or the room and as a way of exploring how to create policy from the bottom up
- *See Mode #5: Distribute Power*

References

"Experimental Modes Convening -- Meeting Notes." *Google Docs*. Accessed September 29, 2015. http://bit.ly/modesconvening.

"No More Trickle Down #CivicTech." *Medium*. September 30, 2014. Accessed September 29, 2015. https://medium.com/@elle_mccann no-more-trickle-down-civictech-81341cf48a14.

Stef Milovic of the Hidden Valley Nature Lab and Naheem Morris of the Red Hook Digital Stewards program discuss strategy at the Experimental Modes Convening. April 4, 2105. Photo by Daniel X. O'Neil.

Tools, Not Tech

The textbook definition of "technology" is all about "tools." Not computers, not command lines, but, to quote Wikipedia: "the collection of techniques, methods or processes used in the production of goods or services or in the accomplishment of objectives, such as scientific investigation."

"Civic technologies" are the tools we create to improve public life. To help each other. To make our governments and our communities safe, joyful, equitable places to live our lives.

Over the course of the Experimental Modes project, I focused on how different people create civic technology with their communities—the social strategies and tactics wielded to build tech at the speed of inclusion and make sure the civic problem-solving process is truly collaborative. But what nuts and bolts go into making this work ... work?

At the convening of practitioners, as part of the larger discussion of "civic tech" discussed last chapter, we went around the room and shared two types of technologies (tools!) we use to do what we do. These answers are collected on the table below.

Shifting our understanding of "tech" helps us focus on people. When we stop trying to force specific types of tech solutions and start listening to people for opportunities to take action, we put ourselves in a stronger position for problem-solving. We open up creativity, both in terms of who gets to be creative and how we see what tools are available to us. Some of the best civic tools are the ones we already have in hand, and their "civic" utility is unlocked just by wielding them differently.

As you read through the tool round-up below, ask yourself: what tech do you take for granted that's a part of your civic work?

Two technologies we use in our work

Name	Tech 1	Tech 2
Laurenellen	Email	Cellphone
Maritza	Email	Laptop
Sonja	Cellphones	Video Camera
Whitney	Headphones	Websites
Sanjay	Radio	Google Docs
Allan	Drills	Email
Danielle	Laptop	Phone
Demond	Google docs	Phone
Jennifer	Post-it Notes	Whiteboard
Laura	SMS	Community feedback boxes
Tiana	Blogger	Slack
Jeremy	Whiteboards	Pizza
Stefanie	Social media	Email
Greta	Google Hangouts (love + hate)	Routers
Geoff	Group chat	Collaborative source code wrangling system
Asiaha	SMS	Emails
Marisa	Pen + paper	Adobe Illustrator
Maegan	Flip charts	Markers
Diana	Zines/printing press	Whiteboards
Adam	Story circles	IM
Daniel	Slack	Google chat

References

Wikipedia. Accessed September 29, 2015. *http://en.wikipedia.org/wiki/Technology.*

"Primer for Experimental Modes Meeting." Smart Chicago. April 4, 2015. Accessed September 29, 2015. *http://www.smartchicagocollaborative.org/primer-for-experimental-modes-meeting.*

"Experimental Modes of Civic Engagement in Civic Tech." Smart Chicago. December 12, 2014. Accessed September 29, 2015. *http://www.smartchicagocollaborative.org/work/special-initiatives/deep-dive/experimental-modes-of-civic-engagement-in-civic-tech.*

"Before You Came to This Room." Smart Chicago. April 21, 2015. Accessed September 29, 2015. *http://www.smartchicagocollaborative.org/before-you-came-to-this-room-did-you-think-of-your-work-as-civic-tech.*

"But What Is 'Civic'?." Civic Hall. Accessed September 29, 2015. *http://civichall.org/civicist/what-is-civic.*

Experimental Modes convening attendees using laptops, pens, food, and phones for their work. Photo by Daniel X. O'Neil.

Where Does Community Organizing End and Civic Tech Begin?

What's the difference, when we focus not on labels, but on practice? What can we learn from seeing civic technology not just in terms of products but in terms of process?

At the end of the convening, we reflected on our discussion. I've collected the group's final thoughts and major takeaways below, organized by theme.

Language

The words that we use to describe our work. "Civic tech" is a new term that, while literally descriptive of the work of the practitioners we brought together, doesn't always resonate with these practitioners or the communities they work with. We talked in detail about how the interest in this new idea was destructive ... as well as how it could provide opportunity.

> **Greta Byrum:** *"Think about words like 'disruption': it captures the interest in short term impact, but it has this problem of not speaking to the long term of real social change and transformation, and it changes our understanding of what work does.*
>
> *Civic tech is the hot new thing. Can we use it in a way that's useful? Can we use it to fuel the work we do? Or will this term undermine the work that we do?"*

Shaping the Narrative and the Practice

Storytelling. Much of our afternoon was focused on questions about documentation: where and how we collect our work and share our models.

Daniel X. O'Neil: *"We're in a sliver of a sliver in the tech space. We need to move from glorifying the anecdotes, the stories we tell to get funding, to sharing the modes and methods and the ways that we do that. That's how revolutions happen, when people share their understandings, when people come together and share with each other the exact ways that we do things."*

Adam Horowitz: *"Where are the stories about the innovations I've heard about today told and how they can be told bigger? We read about Uber in the paper, not about community tech. What's the role of storytellers in making this work more noticeable?"*

Maegan Ortiz: *"I'm thinking about how this tech space was created: who was in the mind of the folks who created it and who wasn't, and how, by using community organizing models, we can either replicate that or we can use it and imagine it and push it to be something different that may even disrupt, interrupt the original vision."*

Community Organizing

More than their use and creation of community technologies, what united the people in the room was their focus on community organizing. What is a collaborative process to make tech if not the collective, organized effort of a group of people looking to make their lives better?

Demond Drummer: *"I'm a tech organizer. I've always had a problem with the distinction between organizing and tech. But from this conversation today, particularly with Maegan (Ortiz), I've come to own and better understand the deliberate, conscious, purposeful use of the "tech organizer" as a tool and a field of play where power itself is contested."*

Diana Nucera: *"It's clear from this gathering of community organizers that we're in a time where community organizing is extremely important in government. So the question is, how do we get government to adopt community organizing? It's always been clear that government should adopt community organizing, but it's now clear there's a need for it. The use of technology has revealed that need. As we go forward from here, I hope we stay true to community organizing practices."*

Comfort and Tension

We talked about "ingredients for engagement": what qualities an organizer instills to not only get people in the door, when it comes time to work together, but to keep them there, make them feel comfortable, and enable an environment where people as individuals and together as a collective can share power and take action. The practices and ideas that came up over and over included "invitation," "permission," "comfort," and "active listening."

On comfort

Sabrina Raaf: *"I keep thinking about how Chicago has this interesting history in the art world of walk-ups and basement galleries traditionally called 'uncomfortable spaces.' I'm struck by the conversations we had today about 'comfort,' and hoping, hoping for a new tradition of 'comfortable spaces.'"*

On tension

Allan Gomez: *"It's important to remember the default settings. The status quo. The default ends up being such an inertia-creating force, it's difficult to change. So I want to semantically challenge the idea of "comfort" because tension needs to be created to change the default. If we're looking for real innovation, we need to look for examples grounded in people's lives from all over the world. Language of reclamation. And we need to reflect on how we want to use this tech versus how this tech forces us to behave."*

Bringing the focus into the immediate present, Tiana Epps-Johnson reflected that even our work in the room that day was an impression of the comfort/tension dynamic:

Tiana Epps-Johnson: *"Comfort in spaces has a lot to do with the people in the room. It's refreshing that a conversation about civic tech is not dominated by white men, and it's not a coincidence that the people who think about community reflect that."*

Expanding on this idea, we discussed that much of our conversation from the day would have been the same if we called it a "community organizers" convening instead of a "community tech" convening, but the people who chose to come (and opt out) would have changed.

Marisa Jahn: *"One of the things that struck me about the different people in the room today is that everyone identifies as a something and something else. Multiple identities. I also have a varied background between advocacy and tech and arts stuff. It's always seemed ad hoc: I used to do things because they interested me or because I wanted to learn or to help people.*

Now I'm thinking about how the way people arrive at tech is through relationships, through connections that validating all the ampersands, all the hats that people wear, all the paths taken."

Many of the Experimental Modes are focused on relationships. Relationships are community fuel and sinew. They are the foundation upon which all community collaboration—tech related or not—is built. Without understanding how social ties work and without investing energy in creating strong, genuine social ties, truly collaborative projects are impossible.

Whitney May, exploring this idea in her own work with local election officials, came up with a formula based on the "ingredients for engagement" discussion earlier in the day: Information + Invitation = Participation.

Information + Invitation = Participation

Whitney May: *"Local government really struggles with reaching out to people, with invitation. And so do we. Our project focuses so much on information, but we need to do more inviting.*

Technology at its best is a way that expands _____. Insert what you will here. For tech to expand community organizing and access to civic information, for me, if I distill that down, it's actually just participation. So how can we use tech to expand participation?"

The answer: we do more inviting.

Jennifer Brandel: *"Information + Invitation = Participation. Thinking about this at a meta level, before I was invited into this conversation about civic tech, I didn't realize I belonged here— or in community organizing. Now I feel like I'm part of something far bigger than I realized."*

References

"Before You Came to This Room." Smart Chicago. April 21, 2015. Accessed September 29, 2015. http://www.smartchicagocollaborative.org/ before-you-came-to-this-room-did-you-think-of-your-work-as-civic-tech.

"Show Your Work: Submit a Civic Tech Case Study." Smart Chicago. April 10, 2015. Accessed September 29, 2015. http://www.smartchicago-collaborative.org/show-your-work-submit-a-civic-tech-case-study.

"Real-world Civic Tech Strategies." Smart Chicago. April 24, 2015. Accessed September 29, 2015. http://www.smartchicagocollaborative.org/ civic-tech-strategies.

Experimental Modes convening attendees looking serious.
Photo by Daniel X. O'Neil.

Closing

When we set out to create civic tools, we hope to change, for the better, our own lives as well as those of the many people who make up our communities, our neighborhoods, our governments, our society. We hope to serve not just "the public good," but people. Real people. People we know and people we'll never know. Our civic goals are thus collective goals, and rooted in good will.

But no matter what our intentions at the onset of our work, what we create is defined by who we are. To take a page from William James: "Our experience is what we agree to attend to. Only those items which we notice shape our minds." Left on our own or only working with like-minded people, we remain limited by our experiences: seeing only the problems we see, building only the solutions we know how to. Though our aim is collective, when we build alone, we build for ourselves alone.

There are whole histories of precedence and good work, some documented earlier in this text and in much more detail in the following Appendix, for alternative practice: methods of doing civic work collectively for collective ends, methods that help us break through the silos of our attention and experience—the false walls that keep us from seeing each other, from listening, and from collaborating as a beloved community. Our hope with the Experimental Modes Project was to begin to illuminate these practices. We believe that if we can attend to existing work being done in the communities we hope to serve, if we can open ourselves to learning from others outside of our expertise, if we can bring more practitioners of civic work together, we can begin to see guidance for doing the work left to do. We can learn how to work better, together.

Civic technology is an emerging field. Its definitions aren't set. Its heroes and leaders are changing. Its scope is vague and its impact hard to define. This is our moment for intentionality in determining what the field becomes. We choose now the practices and values we want to guide us. I hope you'll linger with me at this crossroads before you start walking.

—L(e)

References

"Book Sprints." *Book Sprints. Accessed September 29, 2015. http://www. booksprints.net.*

"Wufoo." *DIY Case Study: Civic Engagement in Civic Tech. Accessed September 29, 2015. https://smartchicago2012.wufoo.com/forms/diy-case-study-civic-engagement-in-civic-tech.*

"Show Your Work: Submit a Civic Tech Case Study." *Smart Chicago. April 10, 2015. Accessed September 29, 2015. http://www.smartchicago-collaborative.org/show-your-work-submit-a-civic-tech-case-study.*

"Active Listening 101 for Civic Tech." *Smart Chicago. April 22, 2015. Accessed September 29, 2015. http://www.smartchicagocollaborative.org/ active-listening-101-for-civic-tech.*

Acknowledgements

Our partners in this project are the attendees of our convening and the organizations we've studied. Here are the attendees of our April 4, 2015 convening:

- Maritza Bandera, On the Table, The Chicago Community Trust

- Jennifer Brandel, Curious City, Hearken

- Aysha Butler, Large Lots, Resident Association of Greater Englewood

- Greta Byrum, Digital Stewards, Open Technology Institute @ New America

- Danielle Coates-Connor, GoBoston2030, Interaction Institute for Social Change

- Demond Drummer, Large Lots, Teamwork Englewood

- Tiana Epps-Johnson, ELECTricity, Center for Technology and Civic Life

- Allan Gomez, Radio Barnraising, The Prometheus Radio Project

- Jeremy Hay, EPANow

- Geoff Hing, News Apps Team, Chicago Tribune

- Adam Horowitz, Imagings, U.S. Department of Arts and Culture

- Marisa Jahn, The NannyVan App, Studio Rev

- Sanjay Jolly, Radio Barnraising, The Prometheus Radio Project

- Sonja Marziano, Civic User Testing Group, Smart Chicago Collaborative

- Anca Matioc, AbreLatAm

- Whitney May, ELECTricity, Center for Technology and Civic Life
- Laurenellen McCann, The Curious Citizens Project, Smart Chicago Collaborative
- Stefanie Milovic, Hidden Valley Nature Lab, New Fairfield Public High School
- Naheem Morris, Red Hook Digital Stewards, Red Hook Initiative
- Diana Nucera, Detroit Future Media, Allied Media Projects
- Daniel X. O'Neil, Smart Chicago Collaborative
- Maegan Ortiz, Mobile Voices, Instituto de Educacion Popular del Sur de California
- Sabrina Raaf, School of Art and Design at University of Illinois at Chicago
- Robert Smith, Red Hook Digital Stewards, Red Hook Initiative
- Laura Walker McDonald, SIMLab

And here's a list of research and other partners: ACTion Alexandria, Art in Praxis, Dodge Foundation, Free Geek Chicago, Free Geek PDX, HomicideWatch, Jersey Shore Hurricane News, LocalWiki, Loomio, New Fairfield Land Trust, Public Workshop, SocialTIC

This project was made possible by two organizations: The John S. and James L. Knight Foundation, led by Alberto Ibargüen, and The Chicago Community Trust, led by Terry Mazany. They provided the funds for this project, and also the space and freedom to let it lead where it may. Knight's commitment to the information needs of communities and the Trust's commitment to being a leader in how community foundations serve those needs formed the bedrock.

Appendix:
Experimental Modes Case
Study Analysis

Introduction

As part of the Experimental Modes of Civic Engagement in Civic Tech initiative, we created the case study sprint, a documentation project to capture, in their own words, how tech practitioners work with their communities in the creation of civic tech. The call was executed in the spirit of booksprints: on a short time frame starting at the Experimental Modes practitioner convening on April 4, 2015. To participate in the sprint, people fill out a Wufoo form, answering a standard series of questions about their project's background, narrative, strategies, and tactics.

What follows is a review of the 17 case studies submitted by April 29, 2015 (16 of which gave us permission to publicly share their content), though the sprint will be left open online for further contribution after this review.

Target group

Although the case study is open to contributions from all, we were particularly invested in documenting projects that pass the People First Criteria:

1. Start with people: Work with the real people and real communities you are part of, represent, and/or are trying to serve

2. Cater to context: Leverage and operate with an informed understanding of the existing social infrastructure and sociopolitical contexts that affect your work

3. Respond to need: Let expressed community ideas, needs, wants, and opportunities drive problem-identification and problem-solving

4. Build for best fit: Develop solutions and tools that are the most useful to the community and most effectively support outcomes and meet needs

5. Prove it: Demonstrate and document that community needs, ideas, skills, and other contributions are substantially integrated into—and drive—the lifecycle of the project

The practitioners at the Experimental Modes convening in Chicago on April 4, 2015 all implemented "People First" projects, so we started there, inviting participants to document their work with us on-site. Afterward, we published an open call on the Smart Chicago blog. Further outreach was conducted through social media and direct asks over email.

Preparation
Sprint participants who also attended the Experimental Modes convening went through an Active Listening exercise in pairs with peers before starting their form responses. Other respondents worked on their own with the form.

Questions
The Experimental Modes team was interested in:

- Language: How practitioners of civic technologies built with, not for their communities describe their work
- Origins: How technology projects that are built with high degrees of community collaboration get started—what prompts this action and who are the action-takers
- Modes: What strategies and tactics practitioners identify using in their work that others can learn from
- Documentation: How and where practitioners already document how they do what they do

Analysis

Overview

We wanted to understand the modes implemented by practitioners in the context of the work that they do, so we started the case study form with a series of questions that illuminated more detail about the who, what, where, when, and why of each project. We invested particular attention to capturing the implementers involved at an individual and organizational level, key details about which technologies were used, created, or remixed as part of the work (passing no judgements or limits as to what tech could be listed), and included questions about how practitioners view their outreach and impact. Here is what we learned:

Project descriptions

The projects documented in the case study sprint ranged from telephone hotlines for domestic worker advocacy to question campaigns that leveraged "glass trucks" and Twitter to enable Boston residents to redesign their transportation system. Some of the case studies focused on tools, like LargeLots.org, others focused on programs, like ELECTricity, and others focused on organizational approaches, such as the work of Radios Populares and FreeGeek Chicago. Common themes found in the "project summary" section include:

Place and space

76% (13) of projects had a clear place-based or offline presence strongly associated with their work.

- San Diego's Open City Project centered on enabling civicly disengaged members of their community to "reimagine" the city.

- GoBoston2030 invites people who "live, work and visit" Boston to contribute to shaping the future of its transportation system "either online or in person"
- Red Hook Wifi and the Red Hook Initiative Digital Stewards Program is specifically targeted to provide and sustain "broadband access in public areas to residents of Red Hook, Brooklyn."0.4167 in

"Public areas" and public spaces were also heavily referenced.

- CUTGroup noted their use of public computer labs for their civic app user testing engagements.
- Hidden Valley Nature Lab connects their place-based learning program to the land preserve adjacent to the high school the project comes from.
- In addition to their specific work, FreeGeek Chicago identifies as "a community space for tech events" which range from Software Freedom Day and a weekly coding group to poetry readings.

Community ownership
35% (6) of projects explicitly and implicitly discussed their project as a tool for different kinds of community ownership.

- The Large Lots program and LargeLots.org enables community ownership of vacant land in their neighborhood.
- The Restart Project and FreeGeek Chicago focus on community and individual ownership of technology (moving from a consume-and-toss tech culture to a culture that fixes, maintains, and owns its tech—or, in the case of FreeGeek that accesses and owns tech potentially for the first time).
- Red Hook Wifi was created for neighborhood control of Internet access and information sharing.

- Radios Populares and EPANow worked with their communities to create media outlets when there were none that served them.

Digital divides

29% (4) projects explicitly or implicitly discussed their work in relation to divides in digital access and literacy, ranging from offering training "rudimentary digital skills" (CUTGroup) to closing civic information gaps for rural, local governments (ELECTricity).

Media

29% (5) projects described themselves explicitly in terms of media and journalism (EPANow, Curious Nation, Radios Populares, Jersey Shore Hurricane News), but several others (including The Nanny-Van App, Red Hook Wifi, and FreeGeek) identified their project as connected to or fueling journalism (i.e. being a news source or offering media training).

Inclusive decision-making

Both the Large Lots program and Curious Nation were specifically oriented toward opening up decision-making (of the allocation of city-owned vacant land and of journalism and editorial choices, respectively), but other projects (including The NannyVan App, ELECTricity, GoBoston2030, and PeerSpring) identified that using methods of open, pro-actively inclusive decision-making in the form of surveys and in-person invitation were core to the direction their project went in.

Geography

Most projects (76%) identified as taking place in a singular city, although there was a good deal of variation. Some identified their geography based on a specific neighborhood (like Red Hook Wifi, located in Red Hook, Brooklyn), while others identified as multineighborhood (like LargeLots.org and the CUTGroup, both based in Chicago). Of the 4 projects that worked in multiple municipalities, 2 identified working specifically at the county-level (ELECTricity and Jersey Shore Hurricane News). The U.S. Department of Arts and Culture was the only organization that identified the locations for their project in terms of "communities," over 150 of which participated in their People's State of the Union project. Additionally, two projects were based in high schools (Hidden Valley Nature Lab and PeerSpring) and two projects were submitted from outside the U.S. (The Restart Project and Radios Populares).

United States
Brentwood, California
East Palo Alto, California
Inyo County, California
San Diego, California
New Fairfield, Connecticut
Washington, D.C.
Miami, Florida
Chicago, Illinois (5)
Takoma Park, Maryland
Boston, Massachusetts (2)
Ann Arbor, Michigan
Detroit, Michigan
Monmouth County, New Jersey
Ocean County, New Jersey
Brooklyn, New York
New York City, New York
Carroll County, Ohio
Hardeman County, Tennessee
Mercer County, Washington
Seattle Washington

International

Mulukuku, Nicaragua

London, United Kingdom

Project dates

(1) - 2003

(1) - 2005

(4) - 2011

(3) - 2012

(1) - 2013

(3) - 2014

(3) - 2015

59% of the projects (10 total) are ongoing.

Practitioners & partners

We asked every participant to identify the individuals who were "key project practitioners/implementers" in their work as well as their key organizational partners. Below, we've grouped the 39 "key practitioner/implementer" roles, categorized by common themes and functions:

21% (8) - executive or founder (i.e. director, executive director, and president)

18% (7) - manager (i.e. coordinator, implementer, manager, program designer)

13% (5) - trainer (i.e. workshop facilitator, codesign facilitator, lead instructor)

10% (4) - techie (i.e. system administrator, developer)

10% (4) - organizer (i.e. community organizer, tech organizer, instigator)

8% (3) - outreach (i.e. dot connector, public engagement, director of community initiatives)

8% (3) - producer (i.e. senior producer, multimedia producer, production planner)

8% (3) - artist (i.e. creator, lead artist)

5% (2) - advisor (i.e. coach, faculty advisor)

5% (2) - public servant (i.e. municipal partner, city lead)

5% (2) - editor (i.e. editor-in-chief, editor)

3% (1) - shop manager

3% (1) - policy wonk

3% (1) - graduate of the program, now involved in running it (i.e. former digital steward)

The number of partner organizations was hugely varied. For GoBoston2030, the Interaction Institute, working with the City of Boston, gathered together over 100 partners. Most others projects surveyed averaged only about 5.

The roles that participants identified "key organizations" playing ranged from funders and incubators to outreach liaisons and co-implementers. Some participants used their answer to this question to name additional individual implementers. Others namechecked specific government agency partners.

Here's a roundup of the most common categories of organizations involved in alphabetical order:

Activist groups/collectives
Foundations
Government departments/agencies
Homeless shelters
Local non-profit organizations
Makerspaces
Meetup groups
National non-profit organizations
Neighborhood associations
Professional associations

Regional planning associations
Schools
Theatre groups
Universities

And here are some of the least common organizations listed, also
in alphabetical order:

Circus groups
Land trusts
Start-up accelerators
Thrift stores

Conditions

In answer to the question, "Was there a particular issue, need, or
desire that brought these players together? If so, tell us a bit about it,"
participants offered a rich array of forces, ranging from natural disas-
ters (the prompt for Jersey Shore Hurricane News) to policy advocacy
(like the Chicago policies that prompted the creation of LargeLots.
org or the opportunity for domestic workers' rights in New York City
that inspired The NannyVan App). The one commonality underlying
every motivation, however, was the need for connection.

In some projects, this need for connection was expressly tied to
technology from the start. The Restart Project was motivated by the
rising tide of electronic waste, paired with tech consumers' increasing detachment from their electronics. CUTGroup was formed to
connect residents and developers on civic projects. PeerSpring and
the Hidden Valley Nature Lab were kickstarted because of both
student and teacher needs to reconfigure the role of tech in the
classroom. ELECTricity was instigated by digital divides in civic
information within rural local governments.

Other projects were motivated by the need to better connect constituents and their governments. GoBoston2030 and San Diego's

Open City Project were government-led initiatives for increasing civic engagement of those residents who don't engage through traditional civic platforms, like voting.

Non-government actors were moved by civic engagement, as well, though this form of connectivity was most often expressed as a way of increasing civic access and/or representation.

- EPANow, a community media project in East Palo Alto, ultimately took root because of the lack of new sources available in the city, generally, let alone serving the age groups of "millennials and those in generation behind."

- Red Hook Wifi was the product of a collaboration between a hyperlocal organization (Red Hook Initiative (RHI)) and a national organization (Open Technology Institute) which partnered to create expand RHI's youth programming to support tech training as a way to supply Internet connectivity in a community that previously had extremely limited access.

- Curious Nation developed out of a journalist's quest to invite the public along for reporting "(yes, physically)" and "share [the] power to assign stories with the public so there weren't only a handful of brains determining the information that thousands of people, sometimes millions, received."

- Radios Populares worked with the Maria Luis Ortiz Cooperative, a women's collective in the rural town of Mulukuku, Nicaragua, to create their own radio station and be their own media in response to the only other radio station in the area spreading misinformation and slander about their work.

The macro view of conditions is pretty well summed up by the U.S. Department of Arts and Culture's one-sentence rationale for why they created their People's State of the Union project: "Democracy is a conversation, not a monologue!"

Inclusion & exclusion

Every case study identified demographics or groups of people that their project did not reach, but that they wish could have. Most also included an explanation as to why.

For some, the identified exclusions were intentional:

- Red Hook Wifi weighed the challenges of age-related programming. "[The Red Hook Initiative, which runs Red Hook Wifi] is a youth development center, so older adults and seniors are not well integrated into the work."

- ELECTricity discussed the limitations of being a pilot project. "We focused on rural areas which means we missed the opportunity to collaborate with election administrators in urban spaces. I think because urban jurisdictions tend to have more resources, especially as it related to technology, the urgency isn't there."

Other projects identified very specific populations, often with context about how they are thinking of resolving their exclusion:

- The Restart Project notes, "We are not always reaching extremely vulnerable groups - we are rather adamant that our community events be open to the general public, so often homeless shelters or centres for people with mental illness are not options because they are often not open to the public. However, we could do a better job of advertising our events nearby to certain vulnerable groups."

- EPANow: "For various reasons we have not yet established a solid connection between a local community health organization, largely because of health privacy laws."

Language barriers were only explicitly cited in only two projects: EPANow and GoBoston2030.

- GoBoston2030: "If we had greater language capacity, we would have reached more people. The website was in Spanish and Chinese, but there are many more languages spoken. Also our organizers spoke English and Haitian Creole, but there are many more languages spoken."

Both the Large Lots program and San Diego's Open City Project discussed complicated issues related to engaging the engaged. For Large Lots, the people that came to the table were homeowners who were already participating in or were predisposed to engage in neighborhood issues. For Open City Project, the problem was just the opposite. "The program ... attracted participants who were dissatisfied [with] the standard pathways of civic engagement. It would have been nice to have the voices of people who make regular use of the pathways in the mix." Similarly, FreeGeek Chicago reflected that while their work was motivated in part as a response to capital flows in non-profit tech, in retrospect, contact with individual donors for unrestricted personal donations would have been a huge boon given the organization's history with "precariously tight budgets."

PeerSpring and Jersey Shore Hurricane News (JSHN) were the only two projects to express relative satisfaction with their reach and little detail about demographics they could extend their work to. PeerSpring cited full ("100%") participation in their classroom experiment. JSHN framed it this way, "JSHN has an enormous reach at the Jersey Shore. It can always reach more, but with the viral nature of news and information, some of the most important stories can reach over a million people in one area."

Technologies involved

To better understand how practitioners create technology for public good with their communities, we wanted to capture an image of all the tools in these practitioners' arsenal.

We asked them to tell us which "technologies were used, remixed, or produced" in the creation of their work, and prompted them to "Think big: Tech ranges from email and wireless networks to 3D printers and MRI machines. Tell us what kinds of tech were critical to this project and **note which ones were created as a result of this project**."

Here's a round-up of all the tech listed, ranked by most common and grouped when redundant, followed by sample explanations of use.

(6) - Email

(5) - Phones

(5) - Surveys (door-to-door, digital, online, question campaign)

(5) - Websites, custom

(5) - Google Apps (Google Drive, Google Bloggers, Google Apps, Google Spreadsheets)

(5) - Software, general (complaint software, PowerPoints, iTunes, graphic design software, VoIP Drupal)

(4) - Computers (reused laptops, recycled laptops)

(4) - Social Media (Facebook, Twitter)

(3) - Flyers

(3) - Internet (wireless networks)

(2) - Radio (broadcast, transmitter)

(2) - Software, custom

(2) - Video Equipment (video cameras, av equipment, video production)

(2) - Coding (open source)

(1) - Audio Equipment (hand-held recorders, mixer, audio editing software)

(1) - Canned Air

(1) - Fab Lab

(1) - Food

(1) - Github

(1) - Isopropyl Alcohol

(1) - Keyboard Shortcuts

(1) - Meetings (village meetings)

(1) - Podcast

(1) - Popular Education

(1) - QR codes

(1) - Rapid Prototyping

(1) - Robo calls

(1) - Screw drivers

(1) - Search Engines

(1) - Story Circles

(1) - Tablets

(1) - Websites (general)

(1) - YouTube

Software, general, GoBoston2030: "The backend of the website is actually complaint software, that's the database for the questions."

Social Media (Facebook), Jersey Shore Hurricane News (JSHN): "I created JSH on Facebook because, simply, that's where my community was gathering: the village square. Why produce news on a blog when I can do it simply on Facebook AND allow 'the crowd' to help report news, share info, help others."

Computers, FreeGeek Chicago: "Recycling and reuse! We help people appropriate technology and break down fear around it by literally breaking it down."

Radio, Radios Populares: "We took basic radio technology that has historically been out of the reach of laypeople and trained people on how to use it, set it up and maintain it. The next key part is the programming and fr [sic] this tech we brought several computers and introduced the community to a basic audio editing program. They got more advanced audio editing software to continue making their radio spot and news programs. The original transmitter was a simple 150 watt transmitter and now they are operating at 300 watts. They used the basic gear and training we provided and have continued to expand on it."

Coding ("open source development practices"), Open City Project, San Diego: "The adoption of best practices from technology fields was more significant than the deployment of tech tools... While the hardware and software projects were completed following Open Source practice, apply the 'anyone can play' Open Source model to designing, building and deploying community improvements is a more substantial intervention."

Food, EPANow: "We eat pizza to keep us going."

QR Codes, Hidden Valley Nature Lab: "The QR code is used to provide easy access to the lesson plans when outside on the land preserve." QR Codes are placed on existing kiosks marking historical sites and nature trails and the students work with teachers, prior to their deployment, to modify existing lessons plans so that they can be accessed and taught while outside on the land preserve.

Traction

In response to a question about how projects got traction ("What guided your outreach/how did people get involved or discover that they could be involved?") participants identified a series of methods very similar to the ones used in network and place-based organizing.

70% (12) of projects discussed practice that involved explicitly meeting their communities where they are in-person. Both EPANow and the Large Lots program reference "doorknocking." GoBoston2030 noted the use of "advertising in subways, buses, local papers and digital billboards". Although flyering was common among many of the in-person advocacy tactics, the CUTGroup went into particular detail of their flyering strategy, which included flyering at all 12 City Colleges of Chicago and an extensive, data-driven flyering campaign at 25 public libraries in areas where their "efforts were lagging". San Diego's Open City Project leveraged an existing "event that was familiar to a segment of [their] target population" and ultimately built their project inside that event because of its communal attractiveness and meaning.

FreeGeek Chicago explained their choice of in-person organizing strategy: "Our outreach was guided by an intense desire to avoid the tech mainstream and go where our audience was: That meant the housing projects, churches, community centers, Stand Down events for homeless veterans, homeless shelters. We assumed correctly that the geeks would show up no matter what we did, and put our efforts into reaching out to poor and marginalized communities."

Some projects set out to create networks, like The Restart Project, which worked through "green groups" to identify community partners who were interested in working on environmental issues related to technology. Others described how their project's outreach was indistinguishable from community work because it was a direct output of existing community activity.

- ELECTricity: "The core of our work is building relationships with the folks who run elections. So, in a way, it felt like we already had traction because 1) the website idea came directly from the community we set out to serve and 2) we had a network for dissemination." (Including a project specific network and a network of project partners.)

- LargeLots.org: "The Large Lots program got traction because 1) it was designed to meet the requirements of local property owners 2) the website made it easier to apply.

- Radios Populares: "We guided our process based on a critique of typical international development work and wanting to address some of the shortcomings inherent in those models.

The process of getting to know the community took longer than usual—it took one year from when we were approached to build the station to when we arrived."

At least 2 projects described utilizing networks that they had previously built through personal and professional experiences, including the U.S. Department of Arts and Culture which maintains a network of "Cultural Agents in 33 sites" and Curious Nation, whose outreach started first with friends and family and then expanded out to reach "a select group of WBEZ sustaining members (aka engaged community members who give money to the station)." (Curious Nation began as a project of WBEZ.)

Both Jersey Shore Hurricane News and Red Hook Wifi received booming attention and community investment as a result of Hurricane Sandy in 2012 and the recovery period after, given that they supplied essential services lacking at that time (news and emergency info on one hand, internet and other communications platforms on the other).

Only 2 of projects explicitly noted the use of social media for outreach at the earliest stage of their project, 4 noted noted the use of email, and 6 described coordination with partners.

Impact

We asked participants to identify how they would "describe the impact of [their] project? (Audience: yourself/your community, not funders.)" The majority of responses focused on the impact of the work on one of three categories:

24% (4) Self or lead organization
18% (3) Individuals
76% (13) Communities

Examples of impact on self

- EPANow: "I've learned so much about launching an organization in someone else's community, and about working with people in a situation of distributed authority where mine is unclear."

- CUTGroup: "Smart Chicago: we have learned a lot about testing, and continue to develop our methodology."

Projects that talked about impact on the self or the lead organization often also explained the impact on the community they were collaborating with.

Examples of impact on individuals

- Red Hook Wifi: "The impact on individuals involved in the [Digital Stewards program, which maintains the Wifi network] is very clear. All participants exhibit growth in their personal or career goals."

- ELECTricity: "We moved local election administrations past uncertainty about tech and in the direction of tech curiosity. The folks we worked with are supremely proud of their new websites." How does ELECTricity measure pride? In the kind of boasting that motivates administrators to help do outreach to communities and drive traffic and attention to their work.

Examples of impact on communities

- LargeLots.org, after reflecting earlier in their documentation about the negative reputation of the Chicago neighborhood of Englewood wrote: "Englewood hit the news once again as a neighborhood where innovative things are happening."

- U.S. Department of Arts and Culture, which hosted a distributed event in 150 communities in the United States shared a micro case study: "After the People's State of the Union, numerous Mexican immigrant and social justice groups who had hosted separate events in San Antonio came together to host a People's State of the Community address on the steps of the Town Hall."

Most projects that described a change in sentiment (The Restart Project: "our volunteers really enjoy our events"; Open City Project: the "process [was] resonant for the larger community", etc), also qualified the limits of their ability to quantify this impact.

The only two projects that noted quantitative expressions of impact in this section were The Restart Project, which notes that they keep data on waste diversion and are "working on a tool to estimate carbon emissions savings" and Curious Nation. On "the metrics that media organizations care about", the stories made with the Curious Nation's first platform (Curious City, hosted at WBEZ in Chicago) out-performed other stories. "Curious City stories comprised nearly 41% of the top 75 stories on wbez.org in 2014, despite only 2% of the stories posted on WBEZ being made with the model. Broadcast wise, Curious City ended up in the top 3 offerings of WBEZ's 50+ shows in an audience benchmarking study, alongside longstanding shows like This American Life and Wait! Wait! Don't Tell Me!"

FreeGeek noted their impact with a single, simple measurement: "We managed to do what we set out to: Create a small but durable institution to connect poor and working class people with technology.

Overview: Modes and Tactics

The Modes and Tactics section of the case study sprint form asks participants to dive "into the nuts and bolts of how your program operates and how you establish[ed] community control." The questions review the 5 Modes of Civic Engagement in Civic Tech (which were derived from observational study of People First tech projects). Participants were then asked to identify which specific modes and tactics they identified wielding and could select as many modes or tactics as they felt were present in their practice. Although the tactics are commonly paired with specific modes, participants could select from the tactics freely, regardless of the modes they identified.

After identifying modes and tactics, participants were then asked to expand on how these practices were used along with a question that asked how else participants talk about their work. Here is what we learned:

What are the key modes (engagement strategies) utilized by your project?

94% (16) - Use Existing Social Structures
76% (13) - Use Existing Tech Infrastructure
76% (13) - Distribute Power
64% (11) - Create Two-Way Educational Environments
53% (9) - Lead From Shared Spaces

Which tactics were most critical to your work? ✷

82% (14) - Partner with Hyperlocal Groups with Intersecting Interests
76% (13) - Leveraged Common Physical Spaces
76% (13) - Be a Participant - Participate in Your Community
71% (12) - Leveraged Existing Knowledge Bases
53% (9) - Treat Volunteers as Members

47% (8) - Remix Tech that People Use for Different End, Don't Invent Something New

41% (7) - Offering Context-Sensitive Incentives for Participation

35% (6) - Use One Tech to Teach Another

35% (6) - Open Up Your Brand and White-label Your Approach

29% (5) - Start with Digital/Media Skills Training

29% (5) - Teach Students to Become Teachers

24% (4) - Paid Organizing Capacity in Existing Community Structures

18% (3) - Co-Construct New Technological Infrastructure

Closer look: Modes

Use Existing Social Structures

This mode refers to tactics for literally meeting people where they are and focus on organizing and relationship-building strategies as critical to the creation of a civic project.

The vast majority (94%) of projects identified that using existing social infrastructure was a primary mode of their work. 13 of the 16 projects who identified using this mode as a primary took the time to share details about how.

6 of these projects identified that their work was nested within existing institutions, such as government agencies (GoBoston2030, Open City Project), media organizations (Curious Nation), schools (Hidden Valley Nature Lab, PeerSpring), and non-profits (Red Hook Wifi, EPANow).

LargeLots.org and Radios Populares operated in a gray space of full affiliation.

- The Tech Organizer that was instrumental to translating the Large Lots Program into the website LargeLots.org was based within a formal institution (the funder of the local planning process), but the work that led to the creation of the civic tool was community-organizing outside of that institution:

connecting existing networks including local civic hack nights and neighborhood associations.

- Radios Populares (RP), in highlighting their work with a Nicaraguan women's collective, identified their presence as part of a greater whole. Although RP partnered with the collective and supplied technical training and collaborative support, it was the partnership that ultimately created the civic tools and their partner who called the shots. The collective "identified who their volunteers and participants would be [and] provided the context and framework for how they wanted us to work/engage them. The process evolved over a year-long getting-to-know period."

CUTGroup, The Restart Project, Open City Project, and The U.S. Department of Arts and Culture all described their use of existing social infrastructure in a catalytic way, noting a strategic vision behind how and why they approach different (formal and informal) institutions.

- U.S. Department of Arts and Culture identifies activating "social justice organizations hungry for cultural programming and creative ways of building social fabric among their constituents." Similarly, The Restart Project approached green groups already working in waste reduction to garner their interest in reducing electronic waste.

- CUTGroup uses the opportunity of test requests to work with new people and identify new populations and community organizations with a stake in or a lot of knowledge about the test subject at hand.

- Open City Project targeted different locations for different phases of their project execution because they were "significant to a specific subset of the program's target audience."

Use Existing Tech Infrastructure

Technical infrastructure refers to both physical elements, like wireless network nodes, radio towers, and computers, as well as digital elements, like social media platforms, email, and blogs—the full range of technical tools a community uses to support everyday activity and public life. This mode focuses on strategies grounded in the reuse, instigation, or remixing of of a community's technological capacity.

Of the 13 projects that documented using this mode, 10 shared more details.

CUTGroup, Radios Populares, Curious Nation, and the Open City Project referenced the use of existing facilities and devices present in their populations.

- CUTGroup allows users to bring their own devices so they can "see how they normally interact with these devices and see if the technology built works for devices that regular residents use everyday."

- Radios Populares was able to use their partner's workspace for training and for the studio; from this start, they were able to bring "forward the capacity for the building of a 150-foot tower."

- Curious Nation is built out of the "existing technologies, skills, and capabilities of local newsrooms."

- Open City Project highlighted the crucial role of their local Fab Lab, "a community center that deploys industrial-grade fabrication and electronics tools to support project-based, hands-on STEM curricula and provide a rapid prototyping platform for local entrepreneurship" which was an essential provider and partner in the work created.

GoBoston2030, the Hidden Valley Nature Project, and Jersey Shore Hurricane News (JSHN) all pointed to their use of social media because, as JSHN put it, "That's where people are."

ELECTricity was the only project to talk about expanding existing digital skills more generally, unrelated to social media or a specific kind of tech.

Distribute Power

This mode describes a series of tactics for coordinating and ensuring that a project's leadership is collective. 13 of the 17 case studies identified using this mode and 10 of those went into more detail.

ELECTricity, Jersey Shore Hurricane News, and the Large Lots Program reviewed how they let participants literally drive the work, from informing tech investment to providing content to directing policy.

Radios Populares and EPANow discussed methods of shared power through methods of consensus and overall project leadership structure.

- Radios Populares: "We devised small group activity to provide more peer-engagement, and also to allow class leadership to be shared in smaller less daunting settings. But also encouraged presentation style share-back from small groups to a larger group."

- EPANow: "We —I and co-director a Future Mashack—just sort of do things that need to be done, often without running them up or through any chain of command, which doesn't really exist at at any rate. These actions could include creating new technology or social accounts, or taking meetings with possible collaborators, or making editorial decisions. Sometimes this leads to crossed signals, but mostly it seems to contribute toward an atmosphere that while slightly confused at times is one where people who like their independence don't feel hampered."

The CUTGroup, the U.S. Department of Arts and Culture, and The Restart Project highlighted methods of sharing their event/program structure so that others could take their model and run with it.

- CUTGroup: "We publish everything on our CUTGroup tests including processes/logistics of testing and the program. We also publish the data from the tests and the key takeaways. The CUTGroup has generated lots of interest from other people interested in civic innovation across the country. Colleagues in Oakland and Chattanooga have started CUTGroup programs there."

Create Two-Way Educational Environments

This mode is focused on the applied practice of active listening and the tactics needed to put this skill into practice. In particular, it addresses the approaches civic tech projects can take to introduce new technical skills and capacities to communities in an equitable manner. Eleven of the 17 projects identified using this mode and 7 shared more details about their practice.

GoBoston2030 and Curious Nation focused their feedback on the role of questions and response in their work.

- Curious Nation: "Reporters learn from the public which stories they'd most like, the public learns what goes into reporting as well as answers to their questions. When question-askers go out with reporters, they learn from one another in action, in the field. More and better questions are asked in the course of reporting."

CUTGroup, EPANow, Red Hook Wifi, Radios Populares and The NannyVan App wield collaborative design ("co-design") practices that are driven by listening.

- CUTGroup: "Testers are being introduced to new apps and websites that they might not have visited before. We are interested in learning how CUTGroup members currently get information (tech or non-tech) and then see how the technology may/may not meet their needs. For example, in a recent

CUTGroup test of the Chicago Public Schools (CPS) website, we wanted to hear from parents how they normally got information about schools and then see if they ever used the CPS website for school information. Another time, we did a test of OpenStreetMap, where some testers edited a map for the first time."

Red Hook Wifi, Radios Populares, and EPANow particularly focused on the role and structure of training that has students guiding their teachers:

- Red Hook Wifi: "The [Digital Stewards] training is supported and taught by area residents. We are able to provide Internet access through Brooklyn Fiber, a hyper local independent ISP. Programs for young adults, and services for them and others were already being provided by RHI. Matriculated [Digital Stewards] teach new cohorts."

- Radios Populares: "[We used] Freirian pedagogy, that is making all the materials relevant to the context that the attendees came from, by providing a scaffolding series of trainings to build upon learning, to engage participants to become teachers of the material and use peer-education to encourage the sharing of learned skills. Every night we reviewed and evaluated the day's accomplishments and established (or adapted) learning goals for the following day. Every day before starting trainings we had a reflection session and also shared the day's goals with everybody."

- EPANow: "We learn from the student reporters and because they are at different levels of skills with regard to video and editing techniques, some take the lead of working with and directing others who are more novice."

Lead From Shared Spaces

Commons are collaboratively owned and maintained spaces that people use for sharing, learning, and hanging out and they are the foundation upon which all community infrastructure (social, technical, etc) is built. In the course of our study, we considered both digital and physical commons. 9 of the 17 projects specified the use of the lead from shared spaces mode and 6 of the 9 went into further detail.

Most projects focused on shared physical spaces:

- The NannyVan App mentioned their focus on New York and "the lived experiences of the domestic workers there."

- GoBoston2030 and the CUTGroup talked about hyperlocal engagement (GoBoston2030's use of "glass trucks" in "neighborhood plazas" and CUTGroup's use of "the library and other public computer centers").

- Red Hook Wifi highlighted that their "work is hosted by area businesses, residents, non-profit organizations and homeowners."

Curious Nation discussed their use of shared space in terms of occupying the spaces in between structures, noting that their work serves as a "conduit between media and their audiences."

Closer look: Tactics

To make it easy on case study sprint participants, in addition to providing links for further reading about the modes and tactics, we also modified the language used to identify the tactics associated with the Experimental Modes so they could stand-alone as concepts. Participants were invited to share how they used at least 2 of the tactics they identified.

What follows is a closer look at each of the tactics gathered from case study participants, in their own words, along with a few examples from the 5 Modes of Civic Engagement in Civic Tech analysis.

Partner with Hyperlocal Groups with Intersecting Interests

- Radios Populares: "We establish a relationship with local groups to help achieve their goals and also to make sure that we are not imposing our vision of what makes sense. That said it is a through a deliberate dialogue that we get to that type of relationship. The groups we have worked with have had a very strong understanding of what they want to accomplish and in turn we have been able to use that knowledge from them to best understand how what we have to offer will make sense, or perhaps that it or we are not the right fit."

- Hidden Valley Nature Lab: "I partnered with hyperlocal groups such as the New Fairfield Land Trust and the New Fairfield Community Thrift Store. With the former, we both hold the interest of the utilization of the local Hidden Valley Nature Land Preserve. With the latter, we both hold the interest of benefiting the community and, in particular, the public schools system."

Leveraged Common Physical Spaces

- FreeGeek Chicago: "We've grown far beyond just computer recycling because of our (crappy) space."

- CUTGroup: "Public computer centers & libraries as CUTGroup testing locations"

- The Restart Project: "We popped up from the very beginning. We didn't really have office space for the first year we worked, so we always leveraged existing common physical spaces."

- EPANow: "We operate from the facilities of the nonprofit that is sort of incubating us. I say 'sort of' because there was never any formalized arrangement, more a wordless word-of-mouth movement forward into that arrangement. In any event, it has

Experimental Modes

allowed EPANow to very quickly assume a presence within the family of this nonprofit, Live in Peace. That has smoothed the introduction and recruitment of EPANow participants to a point where it is seamless."

Be a Participant - Participate in Your Community

- LargeLots.org, Demond Drummer on his role as a Tech Organizer: "I worked within existing networks: Teamwork Englewood had a deep relationship with the funder of the GHN local planning process and the Large Lots website (LISC Chicago); I was a founding member of the Resident Association of Greater Englewood and friend and advisor to our president Asiaha Butler, who spearheaded the planning of the large lot program; as a longtime member of the open government movement and civic hack night community in Chicago I had good relationships with civic-minded developers. The GHN process created a new relationship with leadership at the Chicago Department of Planning and Development."

- The Restart Project: "We STRONGLY believe in participating in our own communities. So much so, that even after we have other people running Restart Parties across London, we (cofounders) still insist on organising monthly events in our own home communities. It's what keeps us grounded and responding to real needs."

Leveraged Existing Knowledge Bases

- CUTGroup: "Testers share with us their own ways of using tech, their devices, their methods. This builds on how we (Smart Chicago) understand how tech is used. Also, we publish data from every test to build on the knowledge, in hopes of teaching others about what we learned."

- ELECTricity: "The Center for Civic Design developed a re-search-based field guide outlining voters needs when looking for election information online. We collaborated with them to design the look of the website template."

- EPANow: "... Essential to EPANow's development was the recognition that the talent and assets most integral to its success were already in place. In other words, leveraging existing knowledge bases is at the root of what we are doing. That has taken the form of not really seeking outside volunteer assistance, except in the most occasional circumstance and then in a supporting, ancillary role.

  "... We move directly toward the spaces where the expertise and knowledge of people who are participating in EPANow exist: their lives, their homes, their interests, their other activities. For example, stories are generated from the community of EPANow participants, story angles and reporting methods are shaped by their perspectives rather than more "traditional" journalistic approaches. One example under way is that Shana wants to do a story about school violence and bullying. It turns out that because she is in a lot of fights, her suggested approach is to record fights and then interview participants afterwards about what they might have done differently to avoid the confrontation."

Treat Volunteers as Members

- U.S. Department of Arts and Culture: "We create an identity for our participants—'Citizen Artists'—not just volunteers."

- FreeGeek Chicago: "Volunteers aren't just members, they run the show." Literally: FreeGeek noted in their project summary that they are "democratically governed by its volunteer community. The FreeGeek Chicago constitution is built around the concept of small, ad-hoc leadership groups which are accountable to the wider community."

- Jersey Shore Hurricane News considers anyone who submits a photo, an event, a story, or a tip to be a "contributor."

Remix Tech that People Use for Different End, Don't Invent Something New

- The NannyVan App: "Before working on the NannyVan App, we conducted a survey with Domestic Workers United, a New York City-based advocacy group to assess what tools domestic workers had at their disposal. Through the survey, we found that the one thing that all domestic workers have is at least a basic cell. From there, we focused all development on this tool."
- Open City Project: "Before the Open City Project launched, it's design was informed by many conversations with people who were part of either executing ,or planning to execute in the future, similar programs."
- ELECTricity opted to create a website template on Google Blogger rather than creating new websites or even leveraging more a more complex blogging platform because it was a fit for their communities technical and time needs as well as their digital literacy.

Offering Context-Sensitive Incentives for Participation

- CUTGroup: Offers $20 Visa gift cards for testing civic apps.
- FreeGeek Chicago: "Free/cheap computers."

Use One Tech to Teach Another

- The Restart Project: "We believe in dialogic learning—that those who are experts fixing also learn something by working with a participant. And that we need to build from people's existing knowledge and experience. A la Paulo Freire."

Open Up Your Brand and White-label Your Approach

- FreeGeek Chicago: "We took much of our approach and name from FreeGeek Portland, which encouraged groups to use the FreeGeek name for similar programs."

- Curious Nation: "Curious Nation is white-labeling our platform and training so others can employ the same techniques. (E.g. MI Curious.)"

- Open City Project: "This one is still aspirational for the Open City Project, but the plan is to produce a toolkit with separate guides for municipalities, community agencies, and community members to create their own version of our project.

Start with Digital/Media Skills Training

- The NannyVan App: "The NannyVan App was developed after a long period of time establishing a relationship and partnership with the Domestic Workers United through media production trainings."

- ELECTricity: "Before we began work on the election website project, we published quick tech guides on how election officials can use free stuff like Twitter and Survey Monkey to share and collect information with their communities. We developed a survival guide for local election offices, a clearinghouse of free and low-cost resources to help modernize election administration. The website project—using the free Google Blogger platform, Gmail, and Google Drive—was a natural next step."

Teach Students to Become Teachers

- Hidden Valley Nature Lab: "The project truly teaches students to become teachers by literally offering them the opportunity of constructing their own lesson plans to correspond with one's academic curriculum."

- Radios Populares: "We have used this to both encourage sharing of knowledge and to help build upon existing understanding. We often state that the best way to know that you know something is if you can teach it. That said, it is sometimes tricky to teach, so we develop exercises and a process that can be replicated to help organize future workshops, or to involve other members of the community. We also devise trainings that incorporate a shared learning process, including, rotating stations, small group sessions, peer-education, and presentation style information sharing. We feel that this strengthens the capacity to and adds more tools in the toolkit to best be able to carry out a sharing of knowledge once we are gone."

- FreeGeek Chicago: "Students are often better teachers than experts, and we leverage that to the max in our education program."

- Red Hook Wifi noted how graduates of their Digital Stewards program, which maintains the wifi network "recruit and lead new Digital Stewards."

Paid Organizing Capacity in Existing Community Structures

- LargeLots.org, authored by Demond Drummer, Tech Organizer: "I was an employee of an organization within the LISC Neighborhood Network, I was a founding member of the Resident Association of Greater Englewood, where I've settled into an advisory role; and I was a member of the civic tech community in Chicago. As a member of these networks and communities, I was able to craft something resembling a shared win."

Co-Construct New Technological Infrastructure

- Red Hook Wifi is a community wireless network in Brooklyn that is maintained by the Digital Stewards, an educational program for young adults).

- Radios Populares worked with a local women's collective to help them build their own radio station and collaboratively trained the skills needed to keep it up, run programming and, eventually, expand its reach and signal.

- FreeGeek Chicago provides access to free computers built for and by the Chicago communities that need them most.

Closer look: other practices

After completing the above sections, detailing work, use, and context of the Modes of Civic Engagement in Civic Tech, we asked participants to share other ways they talk about their work. Including "strategies or tactics that were important to executing their projects and ensuring community control."

Most responses were either focused on narrative methods and questions (how their work exists conceptually, as a force of power) or the tactical methods at play.

Narrative

EPANow, LargeLots.org, Radios Populares, The Restart Project, and Curious Nation used this space to discuss how the concepts and thinking driving their work.

LargeLots.org and EPANow dug into ideas related to the shaping of their project-specific narrative. LargeLots declared that it "embraced every opportunity to tell [its] story". EPANow weighed the consequences of choosing one frame over another. "There is always the danger that refining [the] story to suit funders will alter the mission and purpose and, most importantly, perhaps, the spirit of EPANow. So we tread that line between being, wanting to be, understanding that change is life, and wanting to stay honest and true to our original vision and desire". Curious Nation identified concepts they use to describe and guide their work, including "journalism as a service," "public-powered journalism," "civic engagement," "co-created stories," "democratization of media," and "making reporters' jobs FUN again."

The Restart Project shared thoughts on how their program relates to the broader context of the tech and maker industries that intersect with it, focusing in particular on gender inclusion. "We believe that if women were not participating as volunteers that we had to create a space for them."

In a similar vein of intentionality, Radios Populares offered more context on the solidarity framing used in their work. "The solidarity aspect keeps us rooted to the two way street of developing transformative change. We bring something but also get something in the process. Our mutual liberation is intertwined. This is key to avoid having a hierarchical frame of mind when working with communities. It also helps to avoid the paternalistic/charity tendencies that occur when a "funded" group comes to work with an economically depressed community."

Tactical

Use community events to meet people where they are: Both Hidden Valley Nature Lab and Open City Project namechecked the value and importance of connecting their work to, as Open City Project put it "another that has a wider audience and it's own communication strategy to an overlapping audience". For Hidden Valley Nature Lab, the particular event was a local Earth Day festival. For Open City Project, it was the National Day of Civic Hacking.

Keep an open door: Several projects discussed specific tactics to maintaining low barriers to entry.

- The NannyVan App explicitly cites this as the reason they often call their work "public art." "By referring to these nanny hotlines as 'public art' the topic became more approachable: domestic workers were more excited to participate, and referring to their contribution as "art" valorized their creative agency. For employers, framing the project as art dismantled their inhibitions and allowed them to explore the topic in a new way."

- Open City Project cites their emphasis on participants creating not products but "big tent" ideation projects as way to keep the barrier to participation as low as possible.

- For FreeGeek Chicago, keeping an open door was both literal and the framing they used to explain why they avoided mainstream tech communities. "We want to be a space where honest people can get a break—whether they've got a felony under their belt, don't have their citizenship papers, don't have a traditional gender identity, are considered too old for technology, have mental illness or addiction issues, whatever. Lots of people are invisible to the mainstream technology world, or scary for whatever reason."

Build with, not for: Radios Populares wrote about this idea as "build together," and it was echoed in the comments from ELECTricity, The NannyVan App, FreeGeek, and Red Hook Wifi, too.

- "Whenever possible," Red Hook Wifi noted, "[Digital Stewards] determine the progress and development of the program. We have many, many partners, both in Red Hook and in surrounding areas that support and enable the work. The work is guided by the needs of public housing residents."

- ELECTricity contextualized this aspect of collaborative control in their work with local governments. "It was...important for us to not do the work for the election administrators. For example, Kat and Marie wanted to embed PDF sample ballots on their website. I could have quickly done this, however, what would happen in 6 months or 12 months when they needed to embed PDF sample ballots again? Instead of doing it for them, I wrote step-by-step instructions and created a practice website for them to play in. This approach gives them control and creates sustainability."

- The NannyVan App called this idea "taking leadership from the most impacted," and described how "both New York and national nanny hotlines were created...working in concert with local domestic worker leaders, REV- involved their participation through storytelling workshops that include voice acting, skits, drawing, and envisioning ways to tell a critical [information] in a compelling and creative way."

- FreeGreek Chicago used the framework of straight-up "grassroots organizing": "I can't even say we went to where there were needs — all the founders were already there, and we just built outward. While privileged enough to mostly have college educations, we were all rather poor and we were all working in poor communities."

Radios Populares encapsulated "build with" tactics this way: when we build with, even more than ensuring that communities have a behind-the-scenes understanding of how technology works, "they have a sense of investment in the process that carries them to a greater sense of ownership in the technology overall. When it comes to having a sustainable future for any development project this last aspect is key."

Overview: other documentation

In addition to the specific strategy questions, we were curious whether and how practitioners shared their work, as well as whether they maintained an online presence for this work.

Does this project/program currently have a home online?

The vast majority (94%) of projects captured have an existing digital presence, either within their organizations or as a stand-alone project website. (88% (15) of those with an online presence indicated project-specific webpages or social media sites.)

Have you documented this project's strategies or tactics before?

- 53% (9) projects indicated "yes" when asked if their project's strategies had been previously catalogued and shared links to content from their own work as well as media or academic documentation.

- 41% (7) projects said that their work had "sort of" been catalogued online.

- 6% (1) project said definitively that their work had not been catalogued online.

Would you be willing to share any documentation that's not online?

- 82% (14) of projects indicated that "yeah!" they would be willing or able to share documentation that is not online.

- 18% (3) projects said "no," they would not be willing or able to share documentation that was not currently online. 2 of the 3 declined to indicate why. The third, PeerSpring, indicated it was because this exercise was their first time documenting.

Results

The projects represented in this study contrast significantly with the commonly represented image of "civic tech" as a software and code-dominated arena. Although some of the projects, like Curious Nation, do have a substantial software component, the vast majority represent a great diversity of tech: consider GoBoston2030's complaints software and glass trucks or the community-owned radio station that Radio Populares helped build in Nicaragua. Further, while most of the dominant narrative of "civic tech" focuses on the creation of new tech, the tech projects in this study were often the product of appropriating or reimagining existing tools, ranging from Facebook (home to Jersey Shore Hurricane News) to the combination of email, YouTube, and Google Spreadsheets that became U.S. Department of Art and Culture's People State of the Union, a collaborative storytelling platform for civic engagement.

This deviation could be the product of a number of different influences imprinted from the messaging associated with this project, ranging from the outreach strategy and timeframe of the sprint to the overall Experimental Modes project frame and the standard set by the People First Criteria for submission. Still, the richness and variances present here are critical to reflect on as we evaluate not just the particular results present, but what lessons we take from them.

After review of these case studies and further consideration of the questions (noted earlier) that got us started, here's what we learned:

Language

We learned from this exercise that when strongly tied to place, case study participants use language that is strongly tied place (present in 76% of case studies' summaries) and community ownership (35% of case studies summaries). Although on an individual project basis there was much variation in the specific words used to describe activity, case studies consistently returned to language like "engagement," "civic participation," "organizing," and often included narrative about how the inclusion and exclusion of different communities and community members were part of the conditions that spurred their project's creation.

This theme of organizing was clearly outlined in the final tactical section where projects shared the methodologies and thinking that influence their work. Both The Restart Project and Radios Populares directly referenced education frameworks advocated for by Paulo Freire.

EPANow wrestled directly with questions of identity and language framing, highlighting the need to "code-switch" or reframe the work that they do depending on who they're talking to in the community or outside the community (i.e. funders).

Further study is needed to determine if "civic tech" and/or "community tech" are useful frameworks for understanding the arena of "tech for public good" and whether or not these terms resonate with the ways real practitioners of these technologies understand, teach, and do their work.

Origins: The projects in this study got off the ground and got traction in response to one of two forces (either stand alone or in combination): immediate or systemic need for connection.

Immediate need for connection is generated by sudden circumstance, such as when a natural disaster strikes (like Hurricane Sandy, which initiated the creation of Jersey Shore Hurricane News) or when a major policy drops (such as the historic domestic worker's law passed in New York that instigated the creation of The Nanny-Van app). In order to respond, action needs to be taken quickly, which can only be done and maintain a community-led focus if the project leads have deep pre-existing relationships (like the partner groups involved in the Large Lots Program and Red Hook Wiki) with their communities.

Systemic needs are those that evolve over time, with one group of people being continually disadvantaged by the way things are. This disadvantagement may be caused by accident or on purpose, and though these needs present no rushing fire to put out, their existence and the desire to resolve inequity associated with them makes systemic needs a major motivator for civic projects. At least 5 of the projects reviewed in this case study analysis were formed specifically to address inequity in communal connections to tech through means like digital literacy training and electronics refurbishment. Several projects, including the two largely steered by government actors, looked at inequities of representation in civic engagement at both a general and issue-specific level.

Partnerships and social connection played an incredibly consistent role throughout every case study. (Explored in more detail below.) Perhaps the biggest takeaway here is the simplest: civic projects that are community-led inherently can't be done alone.

Modes

Form questions about the modes were present to (1) measure whether these methods accurately capture the way practitioners work and (2) learn which modes were most commonly used in the creation of these community-led tech projects.

We learned that, overwhelmingly, case study participants identify using "Existing Social Infrastructure" as key to their work. 16 of the 17 projects (94%) selected this mode. 82% (14 of 17) also selected the primary tactic that accompanies it: Partner with Hyperlocal Groups with Intersecting Interests.

This emphasis on vesting work not just abstractly in communities but concretely in community partners was also reflected in the second-most popular tactic "Be a Participant—Participate in Your Community," wielded by 76% of participants.

These three approaches are fundamentally grounded in relational-organizing, a variety of methodologies focused on building

Experimental Modes

strong, genuine social ties with individuals and groups. This kind of organizing is typically associated with place-based activity, which is why it should be no surprise that 76% of participants also identified "Leveraged Common Physical Spaces" as a key tactic to their work. As noted in the case studies, common physical space can be a public square, a library, or the conference room at a local non-profit. The key element is that the relationships and connections you're making are connected to a space where you can be on equal or at least comfortable footing with the individuals you're connecting with. That's the setting for any community-led, collaborative tech project. Methods associated with this thinking (i.e. Distribute Power, Leveraged Existing Knowledge Bases, Treat Volunteers as Members) were also very common (76%, 71%, and 53%, respectively).

In addition, and keeping with the expanded vision of tech described above, the majority of participants (76%) identified utilizing "Existing Tech Skills and Infrastructure" (instead of or in addition to creating new tools) as core to how they worked. We can translate this picture in this way:

Projects documented in this case study invest energy in in-person outreach and build close relationships with individuals as well as communities in spaces they share, often by playing with, discussing, and teaching each other how to get creative with the technology that's already there.

Documentation

53% of the projects captured in this case study indicated "yes" when asked if their projects' methodology had been previously catalogued and shared links to content from their own work as well as media or academic documentation. 41% (7) projects said that their work had "sort of" been catalogued online.

This is in spite of the fact that 94% of projects were able to share links to an existing digital presence, either within their organizations or as a stand-alone project website. (88% (15) of those with an online presence indicated project-specific webpages or social media sites.)

Many projects pull from traditions of education and training that they have adapted to fit their work, but they may not yet have fully documented these adaptations or released them online. When asked if they would share offline materials, the vast majority of participants said yes.

More to come

We think that this case study format is a good tool for stimulating the kind of sharing and documentation we need to spread these modes of meaningful civic engagement. Please continue to add your voice at http://www.smartchicagocollaborative.org/modes.

The Author

Laurenellen McCann (@elle_mccann) is an internationally recognized organizer, tinkerer, and thinkerer based in Washington, DC. A consultant with the Smart Chicago Collaborative and a fellow at the Open Technology Institute at New America, Laurenellen's work focuses on the "civic" in civic technology, innovation, data, and life, more generally, and she writes and speaks often on interdisciplinary practices for building "with, not for" the people you're trying to serve. She also runs The Curious Citizens Project, a culture lab for experimentation with public participation. Previously, Laurenellen founded the Sunlight Foundation's state and local team and served as the director of one of the largest open government community gatherings in the world, TransparencyCamp. She cut her teeth as a community radio journalist and a youth member of her hometown Board of Education. In 2013, TIME Magazine named her one of 30 Under 30 Changing the World. Find her at http://laurenellen.com and http://buildwith.org.

The Editor

Daniel X. O'Neil is Executive Director of the Smart Chicago Collaborative, a civic organization devoted to making lives better in Chicago through technology. Find him at http://danielxoneil.com.

Made in the USA
Coppell, TX
27 November 2019